Liberalizing Global Trade
in Energy Services

Liberalizing Global Trade in Energy Services

Peter C. Evans

The AEI Press
Publisher for the American Enterprise Institute

WASHINGTON, D.C.
2002

To order call toll free 1-800-462-6420 or 1-717-794-3800. For all other inquiries please contact the AEI Press, 1150 Seventeenth Street, N.W., Washington, D.C. 20036 or call 1-800-862-5801.

ISBN 978-0-8447-7163-2
ISBN 0-8447-7163-5

1 3 5 7 9 10 8 6 4 2

The AEI Press
Publisher for the American Enterprise Institute
1150 17th Street, N.W.
Washington, D.C. 20036

Contents

Foreword

In advanced industrial economies, the service sector accounts for a large portion of each nation's gross domestic product. Despite the increasing importance of services trade, the multilateral trading system began establishing rules to open markets in those sectors only in 1995, with the creation of the General Agreement on Trade in Services (GATS) at the conclusion of the Uruguay Round of trade negotiations. Decisions at the end of the round did provide for continuing negotiations in the services area. Only with the renewed commitment to a new round of trade negotiations, undertaken in November 2001 at Doha, Qatar, however, did serious individual sectoral negotiations go into high gear.

The American Enterprise Institute is engaged in a research project to focus on the latest round of trade negotiations on services. The project, mounted in conjunction with the Kennedy School of Government at Harvard, the Brookings Institution, and the U.S. Coalition of Services Industries, entails analyses of individual economic sectors: financial services, accounting, insurance, entertainment and culture, air freight and air cargo, airline passenger service, energy, and electronic commerce. Each study identifies major barriers to trade liberalization in the sector under scrutiny and assesses policy options for trade negotiators and interested private-sector participants.

AEI would like to acknowledge the following donors for their generous support of the trade in services project: American Express Company, American International Group, CIGNA Corporation, Enron Corporation, FedEx Corporation, Mastercard International, the Motion Picture Association of America, and the Mark Twain Institute. I emphasize, however, that the conclusions and recommendations of the individual studies are solely those of the authors.

In this monograph Peter C. Evans analyzes the rapidly evolving world market for energy services, describes the barriers to the international liberalization of the sector, and advances a series of detailed recommendations for new trade and regulatory rules and principles that will allow more open markets for energy services trade and investment. The Uruguay Round featured little discussion or evaluation of energy services despite the fact that the combined services associated with the production of gas, oil, and electricity make up an increasing share of the $2 trillion energy business. Until quite recently state monopolies or private companies with exclusive franchises primarily supplied energy production and services. During the 1980s, however, a combination of technological, economic, and political factors revolutionized the industry. Privatization and liberalization created huge incentives for the outsourcing of services, including oil and gas field services relating to geological mapping, planning for drilling and resource management, and, later, closing down operations, and at the other end of the process, services related to market development, customer relations, and billing. Further, private firms—independent power producers—often took over the construction and management of large-scale power projects. Those firms in turn created new demands for services associated with the development, construction, and operation of such projects, including detailed site selection, demand forecasting, and environmental impact assessments; financial and legal services associated with project financing and the procurement of necessary permits; regulatory negotiations with public agencies; and the provision of ongoing maintenance once the project is launched. Finally, a whole array of transportation and network services became a necessary adjunct to managing whole tanker fleets for oil and gas and integrating transport services with pipelines in the delivery of liquefied natural gas.

Both at the national and at the international level, government policies often lagged far behind the changes in industry structure and competition forced by technological imperatives. Citing a number of studies, Evans demonstrates that liberalization of the energy market can yield large economic benefits to consumers and to national economies. For instance, the introduction of competition in the natural gas pipeline market in the United States resulted in a decline of gas transportation costs of more than $2 billion from 1986 through 1997. A much broader study of some fifty countries, undertaken by the Australian Productivity Commission, estimated that for electricity more efficient regulation (defined by the degree of unbundling of services, private ownership, mandated third-party access to electrical lines, and wholesale trading) reduced electricity prices by as much as 35 percent.

Despite the solid and accumulating evidence of the benefits of both domestic and international liberalization of energy services markets, formal negotiations in the GATS have almost ignored energy as a service sector until quite recently—although that situation has changed dramatically with stepped-up industry pressure on the United States and other governments since the late 1990s. One problem resulted because the initial negotiations in the Uruguay Round did not count energy services as a separate sector, and only a few countries made commitments regarding trade liberalization (primarily in oil field services). Since 2000, however, a core group of countries, led by the United States, Venezuela, and to some degree the European Union, has made a determined effort to clarify and expand the GATS classification system regarding energy services as a prelude to an effort to secure substantial liberalization commitments in the sector by the end of the Doha Round.

Beyond questions of classification Evans argues that negotiations in the Doha Round should aim to

- reduce existing trade barriers and restrain the introduction of new barriers;
- enhance market access and national treatment for energy service providers;
- create a more transparent regulatory environment, particularly for foreign service providers; and
- strive for pro-competitive regulatory reform in WTO member-states.

As with other individual services sectors, trade negotiators in energy services face the issue of how best to incorporate specific commitments especially relevant to creating more competitive international markets for service providers. Although Evans agrees that a case exists for horizontal, across-the-board competition rules because the underlying market failure stems from common factors in many service sectors (natural monopoly, asymmetric information, positive and negative externalities), he nevertheless believes that the unique features of regulatory reform of energy services establish a stronger case for negotiating a separate framework for additional commitments regarding energy regulation. He cites the reference paper that is appended to the GATS telecommunications agreement as a precedent (although the particular issues in that area differ from those in the energy sector). For trade and investment in energy services, the author describes four core areas that need to be addressed to secure pro-competitive regulatory reform: (1) third-party access to essential facilities, including natural gas pipelines, electric power transmission, and, depending on the circumstances, gas storage facilities, oil storage facilities, and liquefied natural gas terminals; (2) market transparency, including real-time access to information on prices, transmission capacity, congestion, and upcoming demand; (3) competition safeguards, particularly regarding potential horizontal market power; and (4) independent regulation through an independent regulatory body to thwart rent-seeking and attempts at political interference by private and public officials and bodies.

In a final section Evans describes additional challenges confronting energy services negotiators in the coming months and years. First, what special provisions and concessions should be made for developing countries? For instance, should development goals allow greater flexibility in market access commitments? Second, how does one balance so-called public service obligations with the liberalization of energy services? For instance, how does one factor in security, environmental, and rural development objectives? Third, should permanent reservations be allowed on scheduled commitments? Should nuclear power be excluded from the negotiations, as the European Union has suggested? Fourth, should emergency safeguards, that is, temporary market closing or investment restrictions, be built into the energy services agreement? And, finally, to what extent should government procurement policies related to energy be included in the negotiations?

That set of policy dilemmas and conundrums poses difficult questions for WTO member-states, but in the end Evans is upbeat on the chances of success for energy services trade and investment by the end of the Doha Round. He concludes:

> Reasons for optimism exist. The era of vertically integrated monopolies with clearly defined service territories and locked-in customer bases is giving way to more flexible market arrangements.... Increasingly, developed and developing countries are recognizing that the benefits of market allocation depend on establishing fair and effective administrative rules and regulations not only in the domestic context but also for international trade....
>
> Folding energy services into the broader Doha agenda has created a new momentum.... Taking advantage of that opportunity to reach a global trade agreement on energy services will help to ensure that developed and developing countries reap the full benefits of more open and competitive energy markets.

CLAUDE E. BARFIELD
American Enterprise Institute

Acknowledgments

I would like to extend my appreciation to Claude Barfield, Carol Balassa, Elbey Borrero, Gary Hufbauder, Aaditya Mattoo, Chris Melly, Brian Petty, Pierre Sauvé, Tatsuya Shinkawa, and Daniel Yergin. This study also benefited from discussions at the Organization for Economic Cooperation and Development–World Bank Meeting of Services Experts, Paris, March 4–5, 2002. In particular I would like to express my gratitude to the delegates from Canada, the European Communities, Japan, Norway, Mexico, Switzerland, Venezuela, the United States, and the United Nations Conference on Trade and Development for their comments and insights. Partial support for research travel was provided by the Carroll L. Wilson Award, MIT Entrepreneurship Center, Massachusetts Institute of Technology.

1
Introduction

The total business turnover of energy products in 2000 approximated $2.4 trillion, making energy the world's largest industrial sector.[1] The industry is often thought of in terms of physical products such as oil, coal, and gas. A wide range of services, however, underpins the production, transport, and distribution of those goods. Those services range from geological mapping of prospective oil and gas fields through trading and marketing of diverse energy products to end-use energy efficiency auditing and energy facilities management. Subcontractors or in-house operations of vertically integrated monopolies once supplied such services. Globalization, privatization, and liberalization of oil, gas, coal, and electricity markets are dramatically changing the structure of the industry and the way that services are delivered. Under more flexible regulatory regimes, energy companies can bundle energy services in innovative ways. Increasingly the process includes international transactions, as firms supply energy services through cross-border trade, through the establishment of local presence in foreign countries, and through the temporary entry of skilled personnel and heavy equipment.

Global trade agreements inadequately treat energy services, despite their growing role in international trade. A major accomplishment of the Uruguay Round was the creation of the General Agreement on Trade in Services. The GATS significantly broadened the coverage of the multilateral trading system by establishing rules and disciplines on policies affecting access to service markets. For two principal reasons, however, the GATS has been less effective than hoped for in supporting the liberalization of trade in energy services. First, the World Trade Organization (WTO) system did not give a discrete classification to energy services. Instead other sector headings subsumed various energy services; in the few cases where they were listed separately, they were defined too narrowly to cover the breadth of energy services activities that emerged as the industry began privatization and liberalization in the mid-1980s. Countries can make commitments about market access that are ambiguous or do not cover the full array of commercial activities now supplied by providers of energy services, particularly those related to upstream energy development, energy networks, and wholesale and retail activities.

Second, few countries made commitments specific to energy services during previous trade rounds. Although thirty-three countries made commitments in mining-related services, which include oil field services, only eight made commitments covering the rather unclear classification entry of "services incidental to energy distribution"; only three made commitments in pipeline transportation of fuels. The relative lack of trade commitments regarding market access, domestic regulation, transparency, and other important disciplines is a concern, given trends in global energy growth and the potential gains from expanded trade in energy services. Estimates by the International Energy Agency indicate that world energy consumption is projected to increase by nearly 60 percent over the next twenty years. More than two-thirds of the increase in demand should occur in developing regions.[2]

The importance of energy as a foundation of economic growth and prosperity calls for WTO members to renew their efforts to remove restrictions on market access and enhance conditions for competition in internationally traded energy services. Progressive liberalization of energy services fits well with the goals of the Doha trade agenda launched in November 2001, which set forth the objectives of new global trade negotiations. Few service sectors have as broad a scope as energy or are as deeply connected with the alleviation of poverty, the diffusion of technology, and the achievement of environmentally sustainable growth.

This monograph examines current efforts to deepen trade commitments regarding energy services. The built-in agenda established under the Uruguay Round increased attention to energy services in 2000. A core group of countries began the process of clarifying the GATS classification system for energy to secure more precise and meaningful trade commitments from WTO members. Now a part of the Doha agenda, the effort could yield significant benefits. Among other improvements, deeper trade commitments in energy among developed and developing countries could

- shine a spotlight on regulatory practices that unnecessarily impede potential gains from open and nondiscriminatory trade in energy services;
- enhance market access and national treatment for providers of energy service;
- create a more transparent regulatory environment for companies providing energy services in foreign markets;
- encourage greater liberalization and pro-competitive regulatory reform needed to promote economic growth and development; and
- facilitate the diffusion of cleaner and more efficient energy technologies and practices.

This analysis begins by reviewing the important role of energy service providers in international trade and then discussing the benefits and barriers to greater liberalization of the energy market. Chapter 4 explores the deficiencies of the current trade regime and the implications of extending GATS rules to the energy sector. The GATS provides governments the opportunity to make additional commitments. Chapter 5 considers the most meaningful additional commitments in the energy sector, including disciplines for third-party access to essential facilities, regulatory transparency, competition safeguards, and independent regulation. Additional issues confronting negotiators in reaching a successful energy services agreement include public service obligations and provisions needed regarding developing countries, emergency safeguards, and government procurement (chapter 6).

The final chapter suggests that the prospects for a comprehensive GATS agreement covering energy services are positive. Such optimism is based on a growing recognition among governments, industry, and consumer interests that international trade rules need to reflect better the competitive transformation in international energy markets and the central role that services now play.

2

The Importance of Energy Services

In the 1980s a combination of economic, political, and technological factors dramatically altered the market arrangements in the energy sector. Through much of the twentieth century, state monopolies or private companies with exclusive franchises primarily supplied energy-related services. Many of those companies were vertically and horizontally integrated and were subject to restrictive regulation. Under such market arrangements, buyers and sellers were constrained in domestic and cross-border trade. Energy tended to be sold on the basis of long-term contracts, with prices that were relatively stable but also opaque along the energy value chain, particularly for gas and power where consumers had little, if any, choice in how they contracted their energy needs. Those conditions not only skewed incentives and limited competition but contributed to sizable economic welfare losses to countries that could not secure the most competitively priced energy to drive their economies.

By the end of the 1990s nearly all member-nations of the Organization for Economic Cooperation and Development and a growing number of developing countries had begun to restructure the energy sector. The initiatives aimed at privatizing some or all state-owned energy companies.[1] Many countries established new market rules designed to increase competition and provide consumers with greater flexibility in meeting their energy needs. The privatization and liberalization created incentives for outsourcing services on a competitive basis. Market players now had incentives to expand trade in energy and abandon fixed long-term contracts in favor of shorter-term contracts linked to spot and futures markets. As energy trading increased and markets became more liquid, more innovative pricing options and financial instruments developed to manage price risk over time. Together, the structural changes increased the international role of energy services to support increasingly competitive physical markets for oil, gas, electricity, and other energy products.

As global demand for energy continues to grow over the next decade and as markets become more open and competitive, the role for energy services will likewise continue to increase. The international supply will be particularly important in oil and gas exploration, power generation, energy trading and marketing, and transportation and transmission networks.

Oil and Gas Field Services

An oil and gas development project generates demand for various services. The exploration process requires services related to geological mapping and prospect evaluation. Recoverable reserves necessitate services to analyze the physical characteristics of the basin as well as financial and legal considerations of development. Actual field development involves further development planning for drilling, reservoir management, and related activities. Detailed engineering and procurement accompany lift systems, terminals, gathering systems, and pipeline infrastructure, followed by services associated with drilling, logging, and testing wells. To maximize production, wells may require special solvent or stream injection services. The even greater complexity of off-shore oil drilling calls for the integration of highly specialized technology and management skills. Depleted fields use a range of services to close down, including plugging wells infrastructure removal and environmental cleanup.

Multinational and state-owned oil companies now out-source most such activities. Oil companies retain responsibilities for the core management of financing and overseeing projects but have gradually relied on field service contractors to supply the rest. The largest service companies provide a full range of oil and gas development services and have global reach. But thousands of smaller oil and gas service firms exist in developed and developing countries.

Firms spend enormous investment outlays on exploration and development annually. In a recent survey, 155 publicly traded oil and gas companies in the United States and abroad spent $81 billion on oil and gas exploration and development in 2000.[2] Much of the investment goes out to service contracts.

Demand for field services will expand with the rising demand for oil. The International Energy Agency expects total world oil demand to grow more than 25 percent over the 2000s, to 97 million barrels per day in 2010 from 76 million barrels a day in 2000 (figure 2-1). At the same time the locus of drilling activity is expected to shift. Oil production in North America and Europe should decline by 2.2 million barrels per day. Increased output from Russia, the Caspian, Latin America, and Africa will more than offset that loss. Thus oil and gas field services will increasingly be in demand outside the United States and Europe.[3]

Independent Power Producers

Liberalized markets have provided new opportunities for private-sector firms to compete in the development and

Figure 2-1 Total World Oil Production, 1980–2010

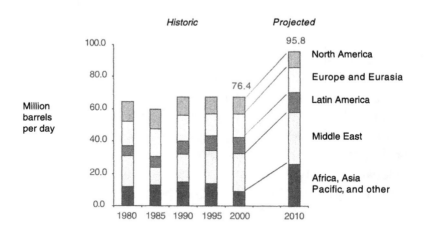

Source: International Energy Agency, *World Energy Outlook* (Paris: IEA/OECD, 1993 and 2000 ed.).

construction of energy infrastructure around the world. One of the most significant changes has occurred in the power industry. Rather than traditional utilities within the confines of their designated service territory, independent power producers (IPPs) now build a great portion of new capacity. In 2000 the twenty largest IPP developers secured final approval for bank financing on 71,000 megawatts of new construction worldwide, up from 8,000 megawatts in 1991. The plants involve an estimated investment of $31 billion concentrated in North America, Europe, Asia Pacific, and Latin America. Depending on market conditions and the progress of liberalization in the power market, IPP developers may build as much as 430 gigawatts of the 771 gigawatts of capacity additions projected in 2000–2010, more than half the worldwide additions to capacity.[4]

The development, construction, and operation of large-scale IPP projects entail a wide variety of services. Initial development involves detailed site selection, power demand forecasting, and environmental impact assessments. The project's legal and financial structures require negotiations with financial backers, equipment suppliers, engineering firms, construction companies, and the network operator. Permitting must comply with federal, provincial, and local government rules and regulations. The project management services often last several years. The construction phase of the project involves a spectrum of design, engineering, and project management services. Once the plant has been commissioned, day-to-day operations must be managed, fuel procured, and preventive and periodic maintenance scheduled and carried out. Aged power plants are rarely abandoned, given the value of developed sites. Instead the existing facilities are usually refurbished with new equipment. The repowering activities involve services such as design, engineering, construction oversight, and management.

Liberalization has not only expanded competition for the construction of large power plants that sell into competitive power pools or to state utility companies. It has also increased opportunities for smaller distributed energy systems. The smaller-scale power-generating technologies are located at or near the source of demand; they typically run on natural gas or fuel oil, although some use wind, biomass, or other renewable power sources. Smaller-scale power systems including plants mounted on barges have alleviated acute power shortages in the Philippines, Brazil, and California. However, distributed energy systems are more typically permanent "captive" installations, relying on direct long-term sales contracts with end-users. In Asia captive power generation makes up an estimated 100 gigawatts of the 900 gigawatts of installed capacity, for a total investment of roughly $100 billion. Although power supplied by small

Table 2-1 Major Energy Trading Markets

Exchange	Location	Year Commenced
Oil		
New York Mercantile Exchange (NYMEX)	New York	1978
International Petroleum Exchange (IPE)	London	1983
Antwerp-Rotterdam-Amsterdam (ARA)	Amsterdam	1985
Singapore Monetary Exchange (Simex)	Singapore	1989
New York Commodity Exchange (Comex)	New York	1992
InterContinental Exchange (ICE)	Atlanta	2001
Natural gas		
New York Mercantile Exchange (NYMEX)	New York	1996
International Petroleum Exchange (IPE)	London	1997
On-the-day Commodity Market (OCM)	London	1999
Electricity		
CAMMESA	Buenos Aires	1992
NordPool (Norway, Sweden, Finland, Denmark)	Oslo	1996
New York Mercantile Exchange (NYMEX)	New York	1996
National Energy Market (NEM)	Australia	1996
New Zealand Energy Market (NZEM)	Auckland	1996
Pennsylvania, New Jersey, Maryland (PJM)	Valley Forge	1997
Amsterdam Power Exchange (APX)	Amsterdam	1999
European Electricity Exchange (EEX)	Frankfurt-Leipzig	2000
New Electricity Trading Arrangement (NETA)[a]	London	2000

a. Formerly the Pool for England and Wales established by the UK Electricity Act of 1990.

plants can be more expensive than power from the central grid, the reliability often compensates for higher costs, particularly where the power supply is unreliable and end-users are sensitive to power disruptions.

Energy Brokers and Marketers

Liberalization of oil, gas, and electric power has greatly expanded opportunities to trade oil, gas, and power in open markets (table 2-1). The ability to trade has opened new opportunities and a need for intermediaries to facilitate transactions between buyers and sellers. Brokers make up one group of intermediaries. After markets are liberalized and barriers to accessing energy networks are reduced, trading volumes tend to rise dramatically. Open-access rules regarding the transmission grid caused third-party sales of electricity to grow to 2.65 billion megawatt-hours (MWh) in 1999 from 0.235 billion MWh in 1996.[5] Each megawatt requires scheduling, price determination, and bill settlement services, which brokers provide for a fee to wholesale buyers and sellers.

Marketers are other intermediaries. They differ from brokers: they take physical positions in the market, usually by building a portfolio of assets, which may include oil fields, gas reserves, gas storage, pipeline capacity, and power plants, or by purchasing and reselling energy products from other providers. Marketers offer a wide range of contract terms that can be tailored to specific customer needs. Energy may be sold on a long-term basis to customers that anticipate a certain level of steady demand. It may also be sold in small allotments in emergency circumstances, when a user is caught short and needs an immediate backup supply. Marketers can add value by their ability to manage price risk with forward physical transactions, forward paper deals, and futures exchanges. They may also add value by serving previously untapped consumer preferences. For instance, some firms offer "green power" to customers that avoid electricity generated from nuclear and fossil fuels in favor of renewable sources.

Transportation and Network Services

The ability to trade energy depends to a great degree on competitive energy-related transportation and network services. Global trade in oil relies heavily on the services supplied by the oil tanker industry. Currently 3,500 vessels of various sizes transport crude oil; each of the 449 largest have the capacity to transport more than 2 million barrels. An active charter market offers both long-term contracts and short-term spot contracts for transport services. Similar markets exist for the international trade in coal. World seaborne trade in coal has grown steadily to reach 480 million tons in 1999, amounting to roughly 13 percent of total world demand.[6] The seaborne trade in liquefied natural gas (LNG) primarily takes place through dedicated ships; some spot cargo transactions have occurred.

Trading in natural gas also depends on pipeline and electric power transmission services. Cross-border trade in electricity and gas is most advanced in North America (between Canada and the United States) and in Europe (between Russia and members of the European Union) and is growing in Latin America (between Venezuela and Brazil for electrical power and between Chile and Argentina for gas). Expanding cross-border trade is generating interest in Africa. Kenya and Uganda are jointly studying a petroleum products pipeline with a power transmission interconnection with Tanzania.[7] Liberalizing third-party access to pipelines and transmission networks opens the market for services associated with such transactions. Pipelines and transmission lines themselves rely on many services to keep them going, including operation, maintenance and repair, and installation and upgrading activities.

Another group of services associated with energy transmission and distribution involves customer metering, billing, and collection, retail services once exclusively provided by utility companies. Liberalization and vertical unbundling of utility monopolies have recently allowed alternative providers to enter the market to supply those services. New market entrants find the ability to contract such services particularly valuable: metering and billing systems are data intensive and expensive to develop and operate.[8] Firms providing billing services can add value by collecting information about customers' energy usage, which can be used to help them understand how usage patterns affect their electric or gas bills.[9]

Integrated Energy Service Companies

Finally, market liberalization is changing how energy services are bundled and delivered. Traditional energy regulation imposed an industry structure that made horizontal integration across oil, gas, and power difficult, if not impossible. By removing those constraints, privatization and regulatory reforms increasingly have made it possible for companies to switch quickly between fuels and bundle energy services in innovative ways. In liberalized markets oil and gas companies own power plants, and electric power utilities have stakes in gas pipelines. Some companies have branched out into other business where synergies can be realized, particularly telecommunications and cable services. In the process some energy companies have gravitated toward asset-based strategies focused on energy manufacturing and energy delivery. Others have gravitated toward energy trading and a focus on energy services.

The reshuffling has partially driven the recent wave of mergers and acquisitions in the energy industry, as companies search to optimize their mix of goods and services. One notable trend has been the convergence between gas and power, with electric utilities in particular investing heavily in gas. In the United States twenty-two mergers of that kind, worth $56 billion, were announced between 1997 and 1999.[10] Similar trends have been taking place in European energy companies. Cross-border mergers and acquisitions have also been on the rise. In the power sector alone those transactions increased from $20 billion in 1996 to $38 billion in 1999.[11]

An increasing number of utility companies, oil companies, and energy equipment manufacturers have sought to build on their respective competencies to supply more energy services on a more fully integrated basis. The companies are now competing to provide a full range of energy asset and energy facilities management services.[12]   Customers benefit by being able to purchase a range of fuel and equipment combined with supporting services from a single supplier. They also benefit from the lower transaction costs associated with one-stop shopping. Companies providing integrated energy services have been particularly successful with large industrial and commercial customers, markets where liberalization is generally the most advanced. Those users care about price, quality, and convenience but are usually indifferent about the type of energy. Paper, steel, chemical, and other large industrial users as well as hospitals, universities, office complexes, and other large commercial users have demonstrated an interest in handing over responsibility for their energy management needs to the energy service companies. In turn companies providing integrated energy services seek to deliver value by improving the energy efficiency of these operations, tailoring energy quality to actual needs, and managing volatility (that is, price risk) over time.

3

Benefits of and Barriers to Liberalizing Energy Markets

L iberalization of the energy market can yield important economic benefits. At the same time domestic regulations often make it difficult for energy service providers to move skilled persons and equipment freely in and out of foreign countries, engage in cross-border trade, and establish a commercial presence in foreign markets. Understanding the benefits and barriers associated with liberalization of this market is central to understanding the value of stronger international trade and investment disciplines.

Benefits of Opening Markets

Liberalization of the energy sector has generally lagged behind other sectors such as airlines, telecommunications, and banking. Even so there are a sufficient number of cases in developed and developing countries to suggest that removing restrictive price regulation, market entry barriers, and other restrictive practices can generate significant economic benefits.

Oil Trade and Competition. Evidence points to significant economic benefits from lifting price controls and phasing out subsidies. U.S. federal regulation of prices for petroleum in the 1970s substantially reduced domestic crude oil production relative to the unregulated environment, and entitlement subsidies resulted in an overconsumption of imported crude oil. Studies suggest that the deadweight losses on the supply and demand sides of the domestic petroleum market were in the range of $1–5 billion per year, not including approximately $1 billion in public- and private-sector costs in administering oil price controls (1980 dollars).[1] The controls also spurred a large transfer of wealth out of the United States. Economists estimate that Saudi Arabia earned approximately $508 billion in additional revenue between 1973 and 1982 because of U.S. price controls on oil. The economic rents enjoyed by Saudi Arabia and other foreign producers vanished almost immediately after oil prices were deregulated in the United States in January 1981.[2]

Reforms adopted by the Japanese government in downstream oil products markets provide another recent example. For years the government restricted the right to import refined petroleum products into Japan to domestic refiners. Those discriminatory rules contributed to widening price differentials between Japan and other industrial countries. The pretax price of gasoline in the early 1990s was triple the U.S. price and two and a half times the European price. The import controls were finally phased out in 1996 with the scraping of the Refined Petroleum Import Law.[3]

Although liberalization has not led to a large volume of refined imports, the threat of import competition was sufficient to increase price competition and erode the margins measured as the difference between retail price and crude oil costs before tax. Between 1996 and 1999 those average monthly margins fell from 29 yen per liter to just 9 yen per liter. Cumulative consumer gains from the reductions amounted to more than 3 trillion yen (figure 3-1). The removal of import barriers also halted wasteful investment in

Figure 3-1 Consumer Gains from Retail Oil Liberalization in Japan, 1996–2000

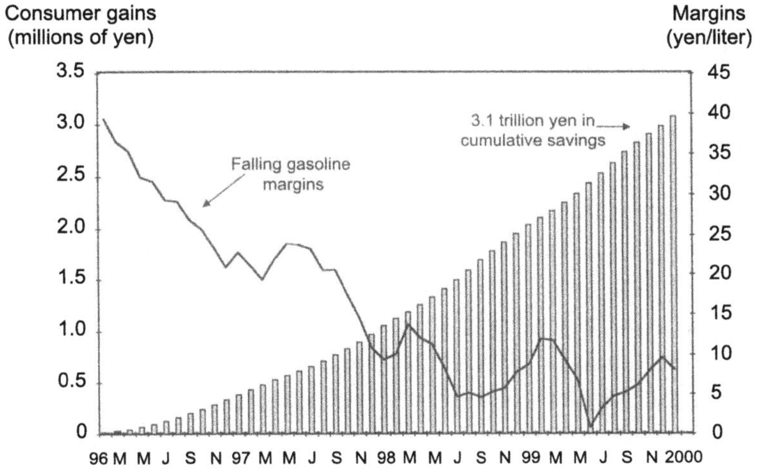

Note: No time lag is assumed between crude costs and retail prices.
Source: For crude costs, Petroleum Association of Japan, Tokyo, 2001; for retail prices, Oil Information Service Center, Tokyo, 2001.

service stations. With scant competition at the retail level, oil companies had a strong incentive to capture the high margins by building more gasoline stations. The number of gasoline stations peaked in Japan at 57,874 stations in 1995. After the market was liberalized, new investment not only stopped, but the number of service stations fell by nearly 6,000 stations by March 2001 as a result of closures and consolidation.

Liberalization of Natural Gas. A substantial body of evidence points to benefits from the liberalization of the markets for natural gas. In the United States following market liberalization in the 1980s, real operating and maintenance expenses in transmission and distribution fell by roughly 35 percent.[4] Pipeline capacity achieved more efficient utilization during peak and off-peak periods after deregulation, primarily as a result of the emergence of secondary markets, which permitted companies holding pipeline space to resell excess space. Those and other improvements in operating efficiency and cost management translated to large gains for consumers. After the Federal Energy Regulatory Commission required pipeline companies to provide unbundled, firm, and interruptible transportation service to other owners of gas supplies, transportation costs to gas buyers declined each year between the major market hubs. Consumers saved almost $200 million per year in gas transportation charges on deliveries to nine major city gates alone. The total decline in transportation charges between 1986 and 1997, based on an extrapolation of results from those nine cities, has been estimated at $2.18 billion.[5] Mandatory unbundling of gas and transportation made gas prices and transportation rates responsive to supply and demand conditions in an emerging competitive national gas market.

In the United Kingdom liberalization of the gas market—which has undergone the longest and most extensive reform in Europe—has yielded positive results. Between 1990 and 1999 average UK industrial gas prices fell by 45 percent while gas prices for other consumers fell by 20 percent. The evidence from other markets is more tentative, in part because liberalization has taken place so recently. Member-states of the European Union adopted major liberalization measures only in 1998. In nearly all EU member-states, gas prices dropped between 1998 and 1999, when oil prices fell to just $10 a barrel, but climbed again when oil prices increased: gas prices remain linked to oil prices. Structural features continue to make it difficult for new entrants to gain access to gas. The supply of gas to Europe remains highly concentrated, with a limited number of upstream suppliers such as Russia's gas monopoly Gazprom. Insufficient gas pipeline infrastructure constrains both access to buyers and sellers and the flexibility needed to trade between markets. Not enough gas hubs have sufficient liquidity to transform the gas market into the kind of commodity market that has emerged in the United States. The EU Commission anticipates that greater competition and lower prices will take hold as prices become more transparent and the linkage of gas prices to oil prices diminishes as gas-on-gas-competition increases.[6]

Liberalization of Electricity. Electricity has proved the most complex and challenging segment of the energy indus-try to liberalize. Despite the problems experienced by California and several other markets, evidence points to overall consumer benefits from the introduction of greater competition. In Europe electricity prices have gone down in all member-states since the EU directive on energy was adopted in 1998, with greater price reductions in markets with the greatest opening of the market. The most signifi-cant price changes have taken place among industrial cus-tomers.[7] UK industrial users have experienced average price reductions of 35 percent. Finland and Sweden prices have fallen by 20 percent and 15 percent, respectively. The drop is notable given that electricity prices in those two countries already were among the lowest in Europe. Major price declines took place in Germany: rates fell by an average of 25 percent between 1998 and 2000. Small enterprises and residential consumers have also gained from liberalization albeit at a reduced rate. Electricity prices have declined in other markets that have opened access, including New Zealand, Argentina, and to a lesser extent Japan.

Just as liberalization has benefits, maintenance of exist-ing regulation incurs costs. Faye Steiner examined the costs of regulation during 1986–1996 for nineteen OECD economies: Australia, Belgium, Canada, Denmark, Finland,

France, Germany, Greece, Ireland, Italy, Japan, the Netherlands, New Zealand, Norway, Portugal, Spain, Sweden, the
United Kingdom, and the United States.[8] Each country's score
tied into key regulatory features including third-party access,
unbundling, and the presence of a wholesale pool while other
factors, including the share of hydropower and nuclear power
in the market, were controlled. Steiner found that the industrial sector benefits most significantly from liberalization
and that expanded third-party access to networks and the
establishment and operation of an electricity spot market
reduced prices.

Australia's Productivity Commission has extended Steiner's
analysis.[9] Looking only at industrial prices, the research
examines the effects of liberalization on fifty economies,
including not only the major OECD countries but also
countries of Eastern Europe, non-OECD APEC (Asia Pacific
Economic Cooperation) members, and countries in South
America. A benchmark of optimal regulation—defined by the
degree of unbundling, private ownership, third-party access,
and wholesale trading—suggests that inefficient regulation
increased electricity prices by as much as 35 percent. The
economies that had most restructured their electricity sectors
had lower industrial prices. The greatest impact on prices was
found in the countries least open to competition, including
Iceland and Switzerland, as well as Turkey, Uruguay,
Venezuela, and Vietnam (table 3-1).

The results of those studies must be treated with care, given
the sensitivity of determining the impact of regulation on
prices relative to the methodology and data employed. As
the authors acknowledge, countries may formally permit
third-party access but block it in actual practice. Neither do
the studies capture the efficiency gains that accrue over time
from better investment decisions. Some analysts consider the
gain in efficiency the most important contribution from

Table 3-1 Estimated Electricity Price Impact of Inefficient Regulation

Range	Percent Increase	Economies (listed alphabetically)
Highest	20% and greater	Iceland, Switzerland, Turkey, Uruguay, and Vietnam
	15–20%	Belgium, Bolivia, Brazil, China, France, Greece, Hong Kong, India, Indonesia, Italy, Korea, Malaysia, Mexico, the Netherlands, the Philippines, Portugal, Russia, Singapore, South Africa, Taiwan, and Thailand
	10–15%	Austria, the Czech Republic, Hungary, Ireland, Luxembourg, Poland, and the Slovak Republic
	5–10%	Canada, Denmark, Germany, Japan, Spain, and the United States
Lowest	0–5%	Argentina

Note: The price impact on industrial electricity rates of inefficient regulation is calculated against a benchmark of optimal regulation based on six variables: unbundling of generation from transmission, third-party access, wholesale pool, ownership, time to liberalization, and time to privatization. Source: Samantha Doove et al., *Price Effects of Regulation: International Air Passenger Transport, Telecommunications and Electricity Supply*, Productivity Commission staff research report, AusInfo, Canberra, October 2001, p. 102.

competition.[10] Not only do the long-run effects take time, however—given the long lead times for construction in the electric power industry—but evaluation requires counterfactuals that are difficult to measure. The limitations notwithstanding, the empirical evidence points to overall gains for countries that have embarked on liberalization, particularly for larger industrial and commercial energy users.

The outlook for further liberalization in the power sector took an abrupt turn as a result of events in 2000 and 2001.

The turmoil in California's power market first raised concerns: the state experienced significant spikes in wholesale prices and numerous rolling blackouts beginning in late 2000 and extending into the summer of 2001. Then in the fall of 2001 Enron—one of the largest energy companies in the United States and one actively involved in competitive electricity and gas markets—collapsed. Sufficient time has elapsed to take stock of what happened and why.[11] The introduction of new supply in the California system played a major role in moderating prices and restoring sufficient reserve capacity to avoid future supply disruptions. In the case of Enron, market players quickly adjusted to its withdrawal from the market, with relatively few effects on U.S. wholesale prices of gas and electricity.[12] Substantial political fallout from Enron's collapse will likely influence the debate over regulation of the energy market for some time. Actual disruption from the company's abrupt withdrawal from wholesale gas and electricity markets, however, has been relatively minor. The limited impact was somewhat surprising, given Enron's once-dominant position as a natural gas pipeline owner, as a commodity trader, as a futures contract trader, and as a marketer, but suggests the relative robustness of the U.S. wholesale gas and power markets and the ability of other firms to fill quickly the position that Enron once held.

Although the market turmoil has shaken support for liberalization, the broad trend to restructure the markets continues. The greatest near-term effect has occurred in the United States. California abandoned retail competition, and at least nine other U.S. states have deferred or canceled plans for retail competition. Spillover effects have shown up in other markets, such as in Canada and Australia, which have slowed reform programs. But U.S. regulators are committed to forming regional transmission organizations to improve the conditions for competitive market access to transmission systems. At a summit in Barcelona in March 2002, leaders of EU member-states reaffirmed commitment to a single electricity market. Japan is expected to introduce further measures to open its electricity market in 2003.

Perhaps most of all, the problems that emerged in 2000 and 2001 sensitized regulators and the industry to the unique complexities associated with liberalizing electricity markets. Greater appreciation exists for the need to put in place workable and properly sequenced restructuring programs, among other things, to manage spot-price volatility, better align supply and demand, and ensure correct pricing signals to users and producers. Liberalization may proceed at a slower pace for a time, but future restructuring is likely to benefit from a clearer understanding of what is needed to produce workable competitive markets.

Barriers to Energy Trade and Investment

The growth in energy services closely correlates with the extent of privatization and liberalization in the energy market. Countries with the most open markets have generally developed the most innovative and cost-competitive energy service providers. Significant barriers, however, remain in most countries. A few examples can illustrate the range of barriers among the different ways in which energy services are provided.

One set of barriers affects the ability of companies to provide services across borders. French companies have relatively unimpeded access to UK and German gas and electricity markets, but the reverse is less true. A similar situation arises between the United States and Canada. Canadian energy companies have significant cross-border supply access to most states in the United States, but U.S. energy companies face market-access restrictions in British

Columbia, Quebec, and other provinces. Another form of cross-border restriction concerns the entry of equipment and tools needed for production or maintenance services. The restriction affects various energy service providers but has been particularly harmful to providers of oil field service, which depend on the ability to move testing equipment, oil rigs, and other specialized equipment from one country to another.

In many cases establishing a local presence in a foreign country is the most efficient and effective way to supply energy services. But companies often face restrictions that can render such efforts noncompetitive or completely disallowed. A common form of restriction concerns foreign ownership. Firms seeking to retain full ownership of their operations may be barred from establishing a local presence unless they join with local joint-venture partners. Rules allowing only minority foreign ownership may restrict mergers and acquisitions. Certain segments of the market may be unavailable to foreign firms. Several countries restrict or completely bar oil companies from engaging in downstream gasoline and other retail marketing. Opaque or discriminatory administrative decisionmaking can create barriers that unfairly disadvantage foreign suppliers, particularly because many energy projects and associated services require extensive licensing or permitting. Finally, in some cases regulators have imposed costly public service obligations on foreign firms that are not required of comparable domestic suppliers.

Countries restrict the temporary entry of skilled persons and managers, typically by imposing unclear or discriminatory rules for multiple-entry visas and the period that managers and other professionals may stay in the country. In some cases temporary entry depends on passing local examinations or other tests before a person is recognized as a professional or specialist. In some cases services provided by self-employed persons are not permitted.

By far the greatest barrier facing international trade in energy services arises from the lack of structural reform,[13] particularly regarding the network-based industries of gas and electricity. Providers of energy services need both nondiscriminatory access to transmission and distribution systems and the right to sell to eligible customers. Brokers of natural gas negotiate with transmission companies for transportation based on their ability to switch their gas from pipeline to pipeline through market hubs to destination. They can provide service only if third-party access is guaranteed and consumer choice has been established. The right to sell to eligible customers has little meaning without access to essential facilities. Likewise the right to access essential facilities is unlikely to yield meaningful competition from new entrants without a clear right to sell to eligible customers.

The terms and conditions under which pipeline and transmission services can be accessed and the degree to which customers are free to choose their preferred supplier vary greatly across countries. Even within Europe significant variation in access terms and customer eligibility exists despite ongoing efforts to form a unified market for gas and power (table 3-2). In the case of electric power, a few countries have granted full choice to retail customers (Finland, Germany, Norway, Sweden, and the UK), while others have followed only the minimum requirements of the European Union directive. A similar pattern has emerged in the case of gas. The gas directive established progressive market opening beginning with a minimum of 20 percent in 2000 and 28 percent in 2003. Although most countries have exceeded those requirements, several—most notably France—have met only the bare minimum of the directive, if that. Similar variation exists regarding the approach to transmission access and the type of vertical

Table 3-2 Gas and Power Liberalization among European Countries

Country	Electric Power Reform Measures		Market Opening in 2000	
	Transmission Grid Access Model	Type of Vertical Unbundling	Gas (%)	Power (%)
Austria	regulated TPA + SB	management	50	32
Belgium	regulated TPA	legal	47	35
Denmark	regulated TPA	legal	30	90
Finland	regulated TPA + pool	ownership	90	100
France	regulated TPA	management	20	30
Germany	negotiated TPA	management	100	100
Greece	negotiated TPA	n.a.	n.a.	30
Ireland	regulated TPA	management	75	30
Italy	regulated TPA + SB	legal	65	35
Luxembourg	regulated TPA	management	51	40
Netherlands	regulated TPA	legal	45	33
Norway	regulated TPA + pool	ownership	n.a.	100
Portugal	regulated TPA + SB	legal	n.a.	30
Spain	regulated TPA + pool	ownership	72	54
Sweden	regulated TPA + pool	ownership	47	100
United Kingdom	regulated TPA + pool	ownership	100	100

Note: TPA = third party access, SB = single buyer, n.a. = not available.
Source: Commission of European Communities, "Completing the Internal Energy Market," commission staff working paper, Brussels SEC (2001) 438, December 3, 2001.

unbundling. In practice some markets in Europe are far more open and competitive than others.[14]

The competitive bottleneck created by monopoly control of Japan's twenty-three LNG terminals provides another example of how the structural features of markets can restrict trade. Japan's vertically integrated power and gas utility monopolies build, own, and operate the terminals, tanks, and regasification equipment associated with the facilities, which receive nearly all the country's gas supplies.

A handful of companies thus control more than 96 percent of the gas available in Japan (only 3 percent is produced domestically) and also the ability to block competitors from accessing those facilities. Given the advantages of gas as a fuel and its importance in power generation, the inability to gain access to Japan's LNG terminals has severely hampered new entry in both the gas and the electricity sectors and has become a point of contention in bilateral trade talks.[15]

Online energy trading platforms are another service tied to the extent of liberalization. Online commodity exchanges for retail energy are attractive because they can pull diffuse players and markets together across a common platform faster and more cost-effectively than traditional approaches to energy sales. Like other exchanges, online energy trading platforms are characterized by great positive network externalities where the power and value of the network grow as the number of participants grows because of increased liquidity and additional options available to users. The ability of the platforms to reach sufficient scale, however, is closely tied to the extent of market reform. Exchanges work only when buyers and sellers have the right to choose and when flexibility in the market allows completing transactions at a reasonable cost.

Unfortunately the cost of barriers to trade and discriminatory practices in energy services is difficult to determine. No economic studies have attempted to estimate the economic costs of those on trade in energy service in any comprehensive manner. Available data on trade in energy services are limited. Statistical reporting among domestic and international energy and trade bodies reflects the legacy of the vertically integrated energy industry and an emphasis on physical flows. Because services were generally bundled within integrated firms—and continue to be in many countries—price transparency continues to be limited along the energy value chain. Still the theoretical case for liberalizing is strong; where it exists, the empirical evidence suggests that much can be gained by opening markets to greater competition. The following chapter examines the issues associated with securing deeper trade commitments from WTO members in energy services.

4

Energy Services and the GATS

core group of WTO members has begun the process of bringing energy services more clearly under the disciplines of the General Agreement on Trade in Services (GATS). By the end of 2001 Canada, Chile, the European Union, Japan, Norway, the United States, and Venezuela had submitted initial negotiating positions. Talks have focused on establishing the scope of issues to be discussed as well as technical issues relating to the classification and definition of energy services. The groundwork, though important, reveals the degree of progress needed before countries—including the many WTO members not yet actively involved in the energy services negotiations—reach a comprehensive agreement.

GATS Modes of Supply

GATS coverage is broader than the trade rules for goods established under the General Agreement on Tariffs and Trade (GATT), which covers only cross-border trade as a means of supply. GATS disciplines include other modes through which services may be delivered.

Three modes of supply relate to energy, as shown in figure 4-1. Mode 1 covers services that are supplied cross-border

29

Figure 4-1 GATS Modes Relevant to Energy

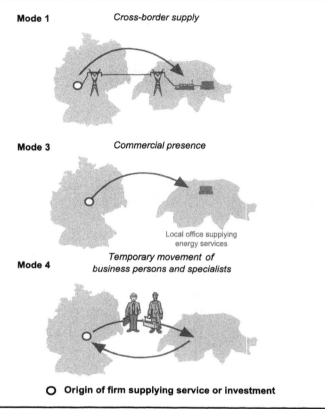

Mode 1   *Cross-border supply*

Mode 3   *Commercial presence*

Local office supplying
energy services

Mode 4   *Temporary movement of
business persons and specialists*

O   **Origin of firm supplying service or investment**

Source: Author and Cambridge Energy Research Associates.

but do not require the physical movement of the supplier or consumer. The category would include cross-border transit or interconnection rights associated with oil and gas pipelines and electric power transmission. Mode 2 has less relevance to energy services. It covers consumption abroad, as with a consumer traveling to the supplying country for services such as tourism and education and consumers traveling to a supply, as well as work such as the repair of aircraft or ships outside an owner's home country.

Mode 3 is highly relevant to energy. It covers services that require the establishment of a local presence for energy services, such as seismic surveying, energy efficiency auditing, and energy marketing, activities that can be reasonably supplied only through the physical establishment of commercial presence in a foreign country. Mode 4 is also relevant to energy in that it covers the entry and visit of those providing services. The category concerns visa terms and conditions, examination requirements, and other regulations that can affect the movement of managers, consultants, and technicians with specialized skills in the course of normal business.

The four modes were developed when the GATS was drafted as a way for countries to organize and schedule their market access and national treatment commitments. Liberalization is more likely achieved if countries make commitments across all relevant modes; they have no obligation, however, to do so. The GATS provides countries the freedom to choose modes in which they will make commitments. They may make commitments in energy services across all relevant modes or selectively choose among them. A country might choose to make commitments on cross-border supply (mode 1) and temporary movement of business persons and specialists (mode 4) but not commit on the right of establishment (mode 3).

General Obligations

The GATS contains more than a dozen general obligations. Two of the most important are most-favored-nation treatment (MFN) and transparency. GATS members must meet those general obligations, with certain caveats and exceptions. The general MFN obligation (article 2) seeks to avoid discrimination among trading partners by requiring that commitments apply equally to services and service providers from all other member-countries. Countries can make exemptions, but those are subject to negotiations and should last no longer than ten years.[1] Thereafter countries must seek a waiver, which must be approved by three-quarters of the WTO members.

The second general obligation is transparency (article 3). No exemptions are permitted; the obligation, however, concerns primarily post hoc notification. Governments must publish all laws, regulations, and administrative guidelines relevant to services trade, and countries must respond to requests from other member-governments to provide regulatory information applicable to the operation of the GATS. Members must notify the WTO on an annual basis of new laws, regulations, and administrative guidelines affecting sectors in which member-countries have specific commitments. In practice, however, the content of those notifications has varied greatly from country to country, with some being quite specific and others general.[2] In its current form the GATS imposes no obligation on countries to consider input from affected parties, including foreign parties, about regulatory measures. It also does not constrain countries from disclosing new or revised regulations at the last minute or excluding affected foreign parties from regulatory deliberations. Creating an across-the-board, or horizontal, discipline to limit those practices and improve transparency through an obligation for prior notification has been proposed; however, it remains unclear whether WTO members will agree to incorporate such a measure into the GATS in the near future.[3]

Two additional general obligations concern domestic regulation and monopoly service providers. In regard to domestic regulation, article 6 states that "each Member shall ensure that all measures of general application affecting trade in services are administered in a reasonable, objective and impartial manner." Members must have judicial or administrative bodies and procedures that provide timely review and appropriate remedies for government decisions affecting trade in services. The article stops short of calling for independent regulation but states simply that WTO members "shall ensure that the procedures in fact provide for an objective and impartial review." All provisions regarding domestic regulation apply only to those sectors for which specific commitments have been made.

Another, if limited, set of obligations concerns monopoly and exclusive suppliers. The GATS permits countries to maintain and even create monopoly service providers but seeks to ensure that such providers do not abuse their market power or compete unfairly by operating beyond the scope of their exclusive rights and thus possibly undermine specific commitments (article 8). Article 9 recognizes that certain business practices may restrain competition and consequently restrict trade. To address that concern, the GATS requires members to consult with one another to eliminate such restrictive practices. No obligations concern the scope and enforcement of policy rules about competition. Equally significant for the energy industry, no general provisions address third-party access to networks or other essential facilities.[4]

Commitments regarding Market Access and National Treatment

In addition to the general obligations that apply to WTO members in all service sectors, the GATS includes provisions for specific commitments. The basic GATS framework lists two: market access (article 16) and national treatment (article 17). Those disciplines come into effect only when they are explicitly listed in a country's schedule of specific commitments, a document appended to the GATS for each WTO member with the member's specific and additional commitments made during or after the Uruguay Round negotiations. The process reflects the so-called positive list approach at the core of the GATS goal of creating more open service markets. Articles 16 and 17 do not apply unless a country has positively affirmed that the sector will be bound by those disciplines.

Energy service providers operating internationally value commitments to market access because they clarify the rights of foreign firms and provide legal standing in a trade dispute. Article 16 lists measures commonly used to restrict market access and asks countries to eliminate these practices. By making a commitment, countries indicate that they will refrain from restrictions on market access that impose limits on (1) the number of service suppliers permitted, (2) the value of transactions or assets, (3) total service output, (4) the number of business persons or specialists who may be employed, (5) measures that restrict or require specific types of legal entity or joint venture through which a service supplier may supply a service, and (6) limitations on the use of foreign capital, such as limits on foreign share-holding or the total value of foreign investment.

Another specific commitment concerns the principle of national treatment, defined as treatment no less favorable than that accorded to similar domestic services and service providers. The commitment would be valuable in trade terms because it would impose an obligation on countries to refrain from maintaining or imposing discriminatory practices that disadvantage foreign service providers. It also establishes a means of recourse to foreign providers of energy services if they are denied equal treatment in the licensing process, taxation, and other regulatory matters.

Few countries made energy-related commitments for market access or national treatment through the GATS during the Uruguay Round.[5] Only three countries—Australia, Hungary, and New Zealand—made commitments in the pipeline transportation of fuels (a subsector of transport services). Eight countries made specific commitments covering services incidental to energy distribution, but several defined the segment narrowly to mean consultancy services. Only two (Australia and the United States) of the eight making commitments were OECD countries. More countries made commitments in services incidental to mining, that is, services supplied on a fee or contract basis in oil and gas fields, including drilling, derrick construction, repair and dismantling services, and casing services. Thirty-three members made commitments in that area, but eleven of those limited their commitments to advisory or consulting services. With its access to the WTO, China recently made limited commitments in that area, but foreign providers of onshore oil field services can operate only in cooperation with the China National Petroleum Corporation (CNPC) in the designated areas approved by the Chinese government.

Other commitments associated with energy deal primarily with construction and retail trade. Forty-six countries made commitments regarding general construction for civil engineering. The area covers long-distance pipelines, communications and power transmission lines, and local pipelines and cables. Thirty countries included energy in their wholesale and retail trade services. Wholesale included trade of solid, liquid, and gaseous fuels and related products; retail included fuel oil, bottled gas, coal, and wood.

Expanding the limited number of specific commitments made during the Uruguay Round would considerably deepen coverage in energy services. The GATS, however, provides members with significant latitude to make exceptions and limitations on specific commitments. Presumably articles 16 and 17 would apply to a sector added to the schedule, but countries may make exceptions if they are clearly spelled out in their schedule (that is, not simply listing a law or measure that contains provisions inconsistent with its trade commitments but indicating the specific provisions that are inconsistent). The process of taking exceptions is sometimes referred to as negative listing. Opportunities to limit actual commitments abound within the GATS framework. Nevertheless the obligation to list exceptions clearly can contribute toward removing trade restrictions by forcing countries to publish discriminatory measures for all to see.

Classification Issues

The WTO services sector classification list (W/120) was developed to help GATS signatories schedule commitments. The classification system, however, does not clearly represent energy services. Energy services were not identified as a separate division when the classification system was devised. At the time state-owned monopolies operating within national or regional markets dominated the energy sector, and oil and gas companies and electric power utilities—whether they were public or private—internally supplied the breadth of energy services activities that emerged since market liberalization. The limitations in the descriptions of energy services found in the UN provisional central product classification mirror the limitations of the W/120. The limitations are problematic because UNCPC is supposed to provide the corresponding central product classification (CPC) number that WTO members use to indicate an offer or commitment in each sector or subsector.[6]

The ambiguity of the classification system impedes the negotiations about energy services. In the three cases where energy services appear in the W/120, the document lists them as part of other generic service entries. Pipeline transportation of fuels is covered as a subsector of transport services. Technical testing and analysis, mining services, maintenance and repair of equipment, and energy distribution services are other business services. When energy services are not explicitly mentioned, where they should fall in the classification system is either uncertain or a matter of dispute.

Recognizing that negotiations on energy services are unlikely to produce meaningful results until the classification system is clarified, Canada, Chile, the European Union, Japan, Norway, the United States, and Venezuela began meeting in October 2001 to rectify the problem. Despite some progress at periodic meetings, the group has yet to resolve several outstanding issues. The task is not simple, given the complexity of the energy industry and its logical overlap with many other sectors. The stakes are also high because the way energy services are classified can influence the terms and content of the subsequent negotiation process. As a result countries have used the clarification exercise as a way to secure their favored negotiation outcome.

At least four main issues must be addressed. One concerns the organization of the W/120. The United States originally proposed creating a separate division within the W/120, with new categories not clearly identified within the CPC. The three existing categories would be moved in their entirety to the new heading. The proposal aimed to create clearer and more commercially relevant listings and to place previously unlisted categories of energy services under the new energy heading. Substantially changing the existing W/120, however, has met resistance from member-countries. Canada has expressed concerns that changes could affect existing commitments. Others have pointed to the potential for additional time and confusion in gaining the acceptance of many WTO members that have not begun to focus on the energy services sector. Still others have argued that even if not specifically identified, energy services are already included since the CPC covers all products and services. Consequently the group's attention has shifted to clarifying where energy service activities may be found within the existing classification structure even though that placement is a less elegant and less user-friendly solution.

Another issue is how broadly to define the scope of energy services. Determining the boundaries of the industry is difficult because energy services are often bundled with other services such as environmental, financial, transportation, legal, engineering, construction, safety, and research and development. One potential solution is to create core and noncore designations, or what is sometimes referred to as a core and cluster approach.[7] The approach would list energy in terms of direct energy services (for example, exploration and extraction) coupled with their associated services (for example, engineering services, environmental services). The method has the advantage of being highly inclusive with less risk that certain types of services would not be covered by the GATS. Conversely the process could complicate the classification system and create duplicate entries. Countries discussing classification issues are leaning toward the creation of a checklist of core and related energy services.

A related issue concerns what sectors are identified as important and relevant to negotiations on energy services. A pertinent example is energy-related shipping services. Given the importance and size of energy-related shipping, some countries, such as Norway, are likely to press for their inclusion in the scope of the talks on energy services. The United States is likely to resist such moves and argue that maritime transport is not relevant to the negotiations on energy services and, if taken up at all, should be part of separate maritime talks. The position reflects strong domestic political pressure to maintain cabotage restrictions, which prohibit the use of non–national flag vessels to transport cargo within the national jurisdiction. The Jones Act restricts waterborne shipments of goods between U.S. ports to ships that are built, owned, and crewed by Americans and therefore prevents foreign flag vessels not only from carrying oil and oil products between U.S. ports but also from serving as transport for offshore oil platforms developed by U.S. companies.

A third issue concerns how detailed to make the classifications. All countries agree that greater detail is necessary to make the classification system more commercially relevant. But countries differ on the level of disaggregation. Japan has proposed general categories whereas Venezuela has been pressing for a high degree of disaggregation, particularly in the area of upstream oil and gas field services. In general an aggregated list tends to facilitate liberalization because it encourages broad commitments, while a disaggregated list makes it easier for countries to omit sectors or list detailed reservations while giving the appearance of committing to many activities.

The last, and perhaps most challenging, classification issue concerns electricity. At least two issues are at stake. One concerns the need to clarify the term *incidental* in the entry "services incidental to energy distribution." It is not clear if commitments based on that entry include electric power generators, brokers, and marketers or only distributors. The original intent of the entry seems to have been those services, such as management, operation, and repair of the network, and meter reading, necessary for the distribution and transmission of electricity on a fee or contract basis.[8] At the time the classification was made, transmission and distribution of electricity itself were rarely undertaken on a fee or contract basis.

The best option may be simply to revise the entry so that it unequivocally includes the actual transmission and distribution of energy, which are now regularly carried out on a fee or contract basis. Since only eight countries made commitments in the area during the Uruguay Round, changing the entry may not be disruptive. Those countries should be permitted to make revisions without penalty if the change expands the country's commitment. Correcting the ambiguity would make future country commitments in the area clearer and reduce the chance that the GATS not recognize important energy services. The time and effort associated with such a clarification seem justified, given the size and importance of the downstream electricity services.

The other issue arises from ambiguity over the definition of electricity. Electricity has the characteristics of both a good and a service. It may be considered a good in the sense that it is manufactured through the process of materially transforming fuels into electrons. It is a service in the sense that it cannot be stored and must be produced as it is consumed. The ambiguity may explain the different way in which electricity has been treated over time. During the first GATT discussions in the late 1940s, negotiators concluded that electricity should not be classified as a commodity. Several countries, however, later took out tariff bindings on electricity and suggested that they considered electricity a good. In a further complication, the WTO secretariat has noted that the World Custom Organization (WCO) harmonized commodity description and coding system (HS) has made electricity an optional heading so that countries are not required to classify it as a commodity for tariff purposes.[9]

The real issue at stake for the GATS negotiations on energy services is how WTO rules will treat the electric power generation sector. As noted, the liberalization of power markets has spawned an international IPP industry, responsible for building an increasing share of additions to electric power plant capacity worldwide. As table 4-1 indicates, which WTO rules apply to most segments of the electric power chain is reasonably clear. Fuels such as coal and oil are considered goods and are therefore subject to GATT rules. With the caveats noted, activities downstream of generation, including transmission and distribution, are services subject to the GATS. But because of ambiguity, electricity produced by IPPs

Table 4-1 Different WTO Rules Potentially Applicable to Power Generation If Electricity Is Classified as a Good or a Service

Classification of Electricity	Fuel	Generation	Transmission	Distribution
Good	Δ	Δ	√	√
Service	Δ	√	√	√
Market share[a]	23.4%	44.5%	8.7%	23.4%

Note: GATT rules = Δ; GATS rules =√.
a. These market share figures are indicative only and vary from country to country and from year to year. The figures presented are for the U.S. power market, which had a total of $218 billion in revenues for fuel, generation, transmission, and distribution in 1998.

could be subject to GATT rules if electricity is considered a good, which implies that it is manufactured. GATT rules apply not to enterprises, only to goods. IPPs could be excluded from market access and national treatment disciplines, which are granted only under the GATS. It could also limit coverage access to essential facilities if WTO members agree to establish such additional commitments.

The implications are considerable, given the significance of the generation sector, which is the largest segment in the electricity supply chain and amounts to almost half of all revenues in the applicable U.S. market (table 4-1). Aside from fuel, generation is the segment of the industry with the greatest potential for competition and has been the most subject to liberalization in recent years. In competitive markets IPPs perform both generation and trading-marketing activities. An IPP could establish two legal entities, one covering the generation business and the other covering its trading-marketing operations. But in practice the two activities are integrally linked. Without the generation function the trading-marketing function cannot be performed, and vice versa. Trade rules should conform to commercial practices. Business should not be forced to establish legal entities to conform to trade rules, particularly where they make little commercial sense.

An odd and less than ideal outcome would result if the structure of a particular power market determined which WTO rules applied to IPPs. Power plants built to serve a single customer—be they captive inside-the-fence plants or build-own-transfer (BOT) projects with a single utility buyer—would fall under GATT rules. An IPP that sought to enter a competitive market with opportunities to sell output to multiple parties, however, would be considered a trader and therefore subject to the GATS. An IPP developer would gain establishment rights (mode 3) and any additional protections such as third-party access rights that WTO members may agree on as part of the GATS negotiations on energy services. An IPP restricted to a single buyer would not have those rights because no investment or network access provisions are associated with GATT rules. Without comprehensive protection for multilateral investment, bringing IPPs under the scope of GATS rules to the extent possible would be preferable to such treatment. Those rules are more encompassing and therefore could give IPPs a greater range of legal protections.

Clarifying classification issues is an important precondition to a successful GATS agreement on energy services. But other issues will also shape the outcome of the negotiations. As governments found with telecommunications, the general obligations and specific commitments contained in the basic GATS framework are not sufficient to reduce trade barriers associated with domestic regulation. To ensure a pro-competitive, transparent, reasonable, and nondiscriminatory regulatory environment requires that WTO members consider developing commitments specific to the energy sector, amended to country schedules as permitted by the GATS. Chapter 5 takes up the nature of those commitments and what they should cover.

5
Necessary Additional Commitments

The most effective strategy for using the GATS to achieve liberalization in services is the subject of debate. The agreements on telecommunications and financial services arose from sector-specific negotiations that had generic elements but also established rules applicable only to those sectors. Some trade negotiators are concerned about relying on that approach in GATS negotiations. Fearing the potentially heavy transaction costs associated with a sector-by-sector approach and creating a confusing patchwork of commitments and obligations, they advocate greater reliance on disciplines that can be applied horizontally to all service sectors. Because the economic case for regulation in all service sectors springs from common underlying market failures (natural monopoly, asymmetric information, and various externalities), generic principles should be available to address those and thereby apply to all service sectors.[1] Proponents of a horizontal approach argue that it can reduce the cost and time associated with international negotiations, can avoid the tendency to focus on politically important sectors at the expense of more encompassing agreements, and can lessen the likelihood that special interests capture sector-specific negotiations.

A horizontal approach, however, may yield less, not more, liberalization. Facing uncertainty, governments rationally act conservatively when making commitments to principles that apply across the board to all service sectors. As a result governments may agree only to horizontal disciplines that are too broad to have much meaning or bite for a specific sector. General principles such as MFN, market access, and national treatment can be powerful tools in the cause of liberalization, but they should be buttressed, where necessary, by rules that reflect the characteristics of a specific sector. The development of specific rules is also more likely to elicit the concentrated effort among regulators, the industry, and other stakeholders necessary to move negotiations forward. Finally a horizontal approach assumes that all service sectors are equally important for economic growth and ripe for negotiation. Both assumptions are questionable. Given energy's fundamental role in driving modern economies, WTO members act reasonably when prioritizing sectors, with energy services high on the list.

GATS article 18 provides a means for countries to negotiate additional commitments not covered by the basic GATS framework. The provision grew from the recognition that MFN, market access, and national treatment disciplines were not necessarily sufficient to ensure the full benefits of trade liberalization. The telecoms' need for additional commitments resulted in a separate telecom reference paper, which set forth additional obligations for WTO members. Creating meaningful disciplines for energy services requires a similar reference paper or annex for energy.

The provisions in the telecom reference paper provide a basis for consideration but do not directly apply to energy. Four core areas are important to securing pro-competitive regulatory reform, some going beyond the principles established for telecoms. The areas are third-party access to essential facilities, market transparency, competition safeguards, and independent regulation.

Third-Party Access to Essential Facilities

The right to interconnect is widely viewed as one of the most important competition safeguards in a network industry.[2] The basic telecommunications agreement would have been far less meaningful without the provisions guaranteeing suppliers access to public telecommunications transport networks or services under nondiscriminatory terms. Establishing the right to interconnect will be no less important for the energy services agreement. But the parallels between telecoms and energy are imperfect. The term *interconnection* and the principles developed to support it in the telecom reference paper are likely to be too restrictive for energy services. The set of principles for energy services must severely limit the ability of a major supplier to refuse access not only to electric power transmission and natural gas pipelines but also to other essential energy infrastructure. Depending on specific circumstances, essential energy infrastructure may include gas storage facilities, liquefied natural gas terminals, oil pipelines, and oil storage facilities.

In developing appropriate language, trade negotiators may look to the essential facility doctrine as it has developed in the context of competition policy in the United States and more recently in Europe. A facility must be shown to have monopoly characteristics that make it truly essential. Suppliers cannot gain access because of inconvenience or some degree of economic loss because of no access: an alternative to the facility must clearly not be feasible.[3] Because competitors can construct oil pipelines and oil storage facilities more readily than infrastructure such as electric power transmission systems, a provision about an essential facility will be less likely to affect them. Reasonable business justifications for denying access may exist; criteria established for that exception, however, should be circumscribed lest it create a major loophole. But when provided, access should be granted in a timely fashion at reasonable fees that reflect the cost of the facilities.

Establishing the basic legal right to third-party access for networks—whether on mandatory or on negotiated terms—is only the first step in ensuring competitive access. Subsequent issues determine the cost, timing, and fairness of connection to a network. One problem in the United States is the gaming among developers to secure the most advantageous place in the interconnection queue. That placement can determine what entity bears the costs of systemwide upgrades, which can range from as little as $100,000 to several million dollars. Another problem concerns what group conducts an interconnection study. Ideally an independent party should undertake the study. But the ideal is often not the case, particularly in markets without an independent system operator or other disinterested party with sufficient knowledge and expertise regarding network conditions. A third access issue is the determination and allocation of interconnection fees. Countries share scant consensus on the cost methodologies that should be employed in determining those charges even though the pricing of interconnection can significantly affect the development of a competitive market. Although the GATS is not the place to resolve all issues, it can encourage governments to establish standardized interconnection policies.

Transparency

GATS-related transparency provisions set forth in article 3 are largely procedural. They require the prompt publication of relevant measures; notification to the WTO of significant changes in laws, regulations, and administrative guidelines;

and establishment of channels for timely responses to information requests from other WTO members. Although the provisions are valuable, they are not sufficient. The transparency disciplines found in the telecoms reference paper and annex offer a starting point. WTO members should also consider adopting a right of prior consultation on draft laws and regulations, with reasonable notice and time for comments.[4] Given the importance of licensing in the energy industry, similar standards should apply to ensure an efficient and fair system for siting, permitting, and construction of new (or retirement of old or inefficient) power plants, pipelines, and other energy-related infrastructure.

But governments should not limit additional commitments to regulatory transparency. And trade negotiators should consider developing language that would focus attention on the need for market transparency. In a competitive context, withholding, delaying, or demanding excessive fees for basic market information can distort competition as readily as physical constraints. As noted, all market participants need access to timely information on prices, transmission capacity, congestion, scheduled volumes, and other data relevant to efficient and fair business transactions. An energy services reference paper would be well served to include provisions that encourage governments to take proactive measures to ensure the free flow of timely information and establish industrywide technical standards.

Provisions to promote market transparency alongside regulatory transparency would have several benefits. The combination would contribute to the goal of improving market efficiency by reducing transaction costs and market distortions. It could reduce the types of questionable energy trading practices that have precipitated regulatory investigations in the United States. It could also reduce the incentives that feed corruption in the energy sector.[5] Creating official

Figure 5-1 Market Share Remains Concentrated in Liberalizing Power Markets

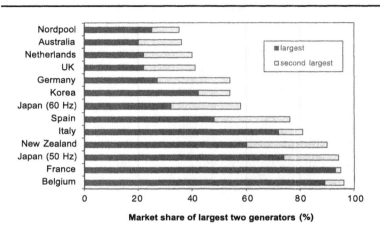

Market share of largest two generators (%)

Note: The United States and Canada are not included, as each is made up of various markets. The share in New Zealand was reduced to 53 percent in 1999. The share in the UK (England and Wales) was reduced to approximately 28 percent in 1999.
Source: International Energy Agency, *Competition in Electricity Markets* (Paris: OECD/IEA, 2001), p. 49.

and transparent channels for providing information to market actors can help reduce existing incentives to bribe officials to gain access to information needed in the course of normal business.

Competition Safeguards

The process of liberalization has drawn attention to two forms of market power. One form is the potential for anticompetitive behavior associated with vertical integration. The incumbent may take advantage of its control over the network (be it pipelines or transmission grid) to favor a more expensive in-house supply with costs (plus a healthy margin) that can be recovered through the regulated business. One approach attempts to control undue market power by policing better the activities of dominant providers. But imperfect information and the regulators' lack of political independence often compromise that approach. Another approach tries to control market power through vertical unbundling. The structural solution is generally considered more effective because it removes many incentives and abilities of incumbent utilities to engage in anticompetitive behavior.

The telecom reference paper offers an avenue for addressing potential abuse by dominant providers. The provisions emerged from the recognized need to prevent major telecom suppliers from engaging in anticompetitive practices, either alone or with others. Specific examples identified in the agreement include (1) anticompetitive cross-subsidization, (2) information obtained from competitors through interconnection negotiations or other means with anticompetitive results; and (3) failure to make available technical information about essential facilities or other commercially relevant information for new entrants to provide their services in a timely fashion.

The provisions directly parallel competition in network-based segments of the energy industry and would greatly strengthen the legal foundation for trade in energy services. The telecom reference paper, however, does not speak directly to cases where a generator or marketer may use its market dominance to control prices. The market share among generators in many power markets remains highly concentrated despite liberalization (figure 5-1).

So-called horizontal market power has been an issue in several power markets. Even without evidence of collusion, studies have shown that market players in concentrated markets may be able to manipulate prices through their bidding behavior even when competitive pools have been established. Studies of the British electricity spot market in the early

1990s, when just three generators controlled much of the market, found that generators were charging prices significantly higher than their observed marginal costs.[6] The giant price spikes experienced in 2000 and 2001 in the wholesale market, particularly in California, have raised the issue of price manipulation by generators and traders.

Perhaps language borrowed from the telecom reference paper could address such anticompetitive outcomes by requiring countries to maintain appropriate measures to prevent major suppliers from engaging in anticompetitive practices. The combination of the vagueness of "appropriate measures" and the specific issues raised by network access, however, weaken the provision so that it could be ineffective in addressing the undue exercise of market power by market actors other than the network operator. Negotiations on energy services must consider whether additional disciplines are needed to address these concerns about competition. Uncertainty over classifying complicates the issue. If generation is not a service, then the GATS may not be the most effective place to seek a remedy.

Independent Regulation

The institutional structures that governments establish to regulate the energy sector vary widely from country to country (table 5-1). In the 1990s privatization and the introduction of competition spurred a general trend toward an independent regulatory agency as the preferred model. Both the EU electricity and gas directives require member-states to establish an independent authority responsible for resolving disputes (article 20). The developments have encouraged the recent establishment of independent regulatory authorities in Argentina, Australia, Belgium, Brazil, Canada, Finland, France, Hungary, Ireland, the Netherlands, Norway, Poland, Portugal, Spain, Sweden, and the United Kingdom.

Table 5-1 Gas and Electric Power Regulation in Selected Countries

Institutional Approach	Countries
Independent regulator and dispute resolution agency	Argentina, Australia, Belgium, Brazil, Canada, Denmark, Finland, France, Greece, Hungary, Ireland, Mexico, Netherlands, Norway, Philippines, Poland, Portugal, Spain, Sweden, the United Kingdom, and the United States
Energy or industry ministry	Austria, China, the Czech Republic, Germany, Indonesia, Japan, New Zealand, Nigeria, Switzerland, Turkey, and South Africa

Source: International Energy Agency, *Competition in Electricity Markets* (Paris: OECD/IEA, 2001), p. 32; and World Bank, "Power and Gas Regulation—Issues and International Experience," draft working paper, World Bank, Washington D.C., April 2001.

Certain characteristics of the energy industry make it particularly susceptible to rent seeking and political interference. A great proportion of assets is sunk, technology exhibits important economies of scale, and customers generally fall into the same grouping as voting populations. As a result, end-user energy pricing has long attracted the interest of politicians. The political sensitivity of prices and the inability of companies to move easily increase the risk of administrative expropriation: the regulators, following public pressure or political expediency, may take actions that push prices below the long-run average cost. The energy industry is full of cases of the struggle between regulatory attempts to extract those quasi-rents and industry attempts to fend them off.[7]

Independent regulation is widely viewed as a way to reduce the problems raised by undue political interference. The OECD and the World Bank advise that policy functions and regulatory functions be separated and that procedures for transparency be enhanced.[8] Decisions isolated from covert pressures are more likely to be made on the basis of facts rather than the influence of government, companies, or other parties. Experience suggests that regulatory decisionmaking can be improved if all communications and evidence submitted to the regulator are made public and if hearings are public and conducted in a fair and impartial manner.

Negotiations on energy services offer an opportunity to reaffirm and codify the importance of independent regulation and transparency in the energy sector. In the case of telecoms, WTO members made independent regulators a requirement but did so without prejudice as to whether the regulator was separate from the ministry that formulates telecom policy. That arrangement should be a minimum requirement in any agreement on energy services.

6

Additional Issues

The process of strengthening GATS disciplines for energy services raises additional issues that negotiators must address. The first concerns developing countries and their special circumstances. Among the questions raised is whether provisions for special treatment and developmental objectives are warranted. A second issue concerns the nature and scope of market restrictions that countries may impose in the pursuit of public policy objectives. A third issue deals with the scope of reservations that countries may take on scheduling commitments. Another issue concerns the advantages and disadvantages of emergency safeguards as part of an agreement on energy services (the GATS does not include safeguard instruments like those in the GATT for goods). Last is the issue of government procurement and whether GATS-specific disciplines should be added in this area.

Considerations of Developing Countries

Developing countries will play a critical role in the outcome of negotiations on energy services. The bulk of the projected increase in world energy demand will take place in developing regions, accounting for approximately two-thirds of the growth in world energy demand between 1997 and 2020 (figure 6-1). Developing regions will become increasingly important buyers of energy services and are likely to

Figure 6-1 World Energy Demand, 1997 and 2020

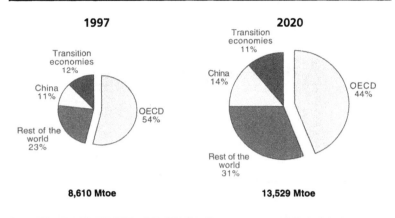

1997

Transition economies 12%

China 11%

Rest of the world 23%

OECD 54%

8,610 Mtoe

2020

Transition economies 11%

China 14%

Rest of the world 31%

OECD 44%

13,529 Mtoe

Note: Mtoe = million tons of oil equivalent.
Source: International Energy Agency, *World Energy Outlook 2000* (Paris: OECD/IEA), p.52.

increase their role as sellers of services, particularly within developing countries markets. Several developing countries have become international suppliers, especially in oil field services. As a result, the scope and benefits of a GATS agreement on energy services will hinge to a great degree on the number of developing countries that agree to make commitments and on the nature of those commitments.

Developing countries present a specific set of issues that WTO members will need to confront. What, if any, special treatment should be accorded developing countries? GATT established a precedent for granting such countries special transitional arrangements in meeting trade commitments and in applying differential treatment for countries at different levels of development. The GATS architecture already offers countries a high degree of flexibility.[1] Any country may impose restrictions on market access as long as it lists them in its schedule of commitments and is bound to provide national treatment only if it explicitly makes such a commitment. The actual need for special treatment calls for careful weighing, particularly for middle-income developing countries, where special treatment might be less warranted, especially if emergency safeguards are incorporated in an agreement on energy services.

Another issue concerns what are broadly called developmental objectives. Developing countries can benefit from establishing linkage programs, which aim to increase domestic sourcing by foreign affiliates and to harness foreign direct investment to upgrading the technology and managerial capability of local firms.[2] What role should the government have in this process? When do developmental objectives justify imposing foreign direct investment and trade restrictions on a discriminatory basis?

Developing countries have sought to preserve the right to place conditions on the market openings that they choose to make in the name of developmental goals. Venezuela's delegation to GATS negotiations has asserted that "the negotiations should respect the developing countries' space to implement policies aimed at domestic capacity-building, in particular the capacity of their small and medium sized energy service suppliers."[3] Performance requirements derive from the view that conditions on trade and investment of multinational enterprises can be used to increase local production, exports, or technology transfer for the host country. To the extent that those requirements impose conditions that are not market driven, however, performance requirements can distort investment decisions and international trade. The trade-related investment measures (TRIMS) agreement reached during the Uruguay Round represented an effort to phase out local content, trade-trade balancing, domestic sales requirement, local hiring targets, and other performance requirements for goods. Developing countries have resisted efforts to extend the reach of the controversial agreement: they often view performance requirements as an important element of their development strategy.[4] The contending views and interests surrounding performance requirements are not easily reconciled and could pose a stumbling block to a GATS agreement on energy services.

A third issue concerns the capacity of developing countries to oversee and regulate competitive energy markets effectively. Although the GATS may promote and consolidate efforts to reform domestic regulation of energy, multilateral rules cannot guarantee sound regulatory institutions. Multilateral trade rules are designed primarily to ensure market access and are not directly intended to promote security, environmental, and social welfare goals. Building credible and independent regulatory institutions requires a parallel effort. The World Bank and other institutions have long supported regulatory capacity building along those lines. Such efforts must continue but must incorporate a greater trade element. Considerable work remains on building compatibility between domestic regulation and multilateral trade disciplines for energy.

Finally, the GATS negotiations need to encourage greater participation by developing countries. Domination by OECD countries in negotiations on energy services would be a mistake, as would a view of the negotiations in North-South terms. The substantial trade in energy between developing countries will only grow. Developing countries stand to benefit not only from establishing stronger multilateral trade disciplines with developed countries but also from establishing stronger trade rules between themselves.

Public Service Obligations

Previous GATS agreements explicitly acknowledge the right of governments to pursue legitimate public policy objectives.

The agreement on financial services concluded in 1997 permits countries to impose prudential measures to protect consumers of financial services and to ensure the overall integrity and stability of the nation's financial system. Maintaining universal service was an important priority for countries making commitments regarding telecoms. The basic telecom agreement and supporting reference paper recognize the right of countries to impose special obligations about universal service.

The nature and the scope of prudential measures will likely loom large in the negotiations on energy services. Many policy objectives associated with a country's energy policy have significant social and economic consequences. In addition to considerations of universal service similar to those with telecoms, health, safety, rural electrification, pollution abatement, energy efficiency, and security goals exist. Governments have invoked all those policy objectives at one time or another to justify quotas, subsidies, and other policy instruments in ways that discriminate against foreign-owned firms. Governments have justified market intervention and the contrivance of competition on the basis of immediate as well as long-term considerations regarding energy security. Immediate security considerations include market interventions taken to ensure the reliability of power systems. Long-term considerations about energy security include market interventions aimed at minimizing a country's dependence on a particular fuel or on a particular country or region.

As uneconomic obligations, environmental, security, and other public policy objectives always present the nagging problem of how they should be met and paid for. Traditionally governments have funded public service obligations by permitting utilities to earn monopolistic rents and practice cross-subsidization. The solution has not been the most efficient. Market liberalization has sought to improve efficiency but has reopened the question of allocating the burden of public service obligations and at the same time treating market incumbents and market entrants fairly.

Although, as often argued, certain market interventions may be legitimate, regulators should strive for symmetry between market players in their application. As Gregory Sidak and Daniel Spulber put it, "Regulators should scrupulously design rules that create no advantage for the entrant over the incumbent, or vice versa, but instead place all competitors on an even regulatory footing."[5] In practice achieving symmetry is not an easy matter. Liberalization often generates strong resistance by incumbent providers because they fear that entrants will cherrypick the market and leave them saddled with the costs of meeting various social objectives. In the context of the GATS negotiations, those concerns help to explain why some incumbent utilities have already expressed reservations about establishing pro-competitive trade rules for energy services.[6]

For those reasons, the process of introducing competition while ensuring symmetry between market players often requires new institutional arrangements and cost-allocation mechanisms. One example is renewable portfolio standards (RPS) for electric power generation[7] to create incentives for inducing higher levels of renewable energy technologies than supplied in a purely competitive context. Establishing certain mandatory portfolio requirements (numerical quotas) obliges electricity suppliers to produce a certain percentage of electricity generated from renewable energy. Governments may also issue certificates for the amount of renewable electricity generated, which can be traded in secondary markets. Producers that do not meet the minimum standard or hold a sufficient number of certificates face penalties. If applied fairly and reasonably, the approach can achieve the public goal of promoting certain energy sources cost-effectively.

Thus, the issue for trade negotiators is not whether governments have a right to pursue legitimate public policy objectives, but how these goals are achieved. Multilateral rules should ideally make it difficult for governments to rely on trade restrictions to achieve public policy objectives. In most cases those objectives are better achieved through other, nondiscriminatory means.[8]

Reservations on Scheduled Commitments

Another issue facing negotiators is what areas to exempt from GATS disciplines. Some existing proposals would exclude certain areas from the talks. The United States and Venezuela have proposed excluding the ownership of publicly owned natural resources.[9] The initial proposal by the EU argued the necessity of such an exemption for nuclear power. International nuclear trade is subject to preexisting international agreements or specific provisions in more general agreements and therefore cannot be assimilated to general energy trade.[10]

Given the delicate nature of national sovereignty involved in national resources and the safety and security concerns associated with nuclear power, those proposals may be inevitable and politically expedient. But accepting those exclusions in their entirety and without careful review does carry a cost. For natural resources such exclusion could permit governments to continue to discriminate against foreign firms in leasing blocs for oil exploration and thus exclude many energy services that can be supplied competitively. Excluding nuclear power has similar implications. Service opportunities relating to hazardous waste management and decommissioning of aged reactors will expand. By potentially limiting competition, exclusion could slow the diffusion of new technologies and management techniques. Negotiators should consider blanket exclusions carefully.

Emergency Safeguards

GATS article 10 calls for negotiations on emergency safeguards, but the current agreement does not include such provisions. Some analysts have argued that the economic case for safeguard measures in services is weak. Bernard Hoekman and Michel Kostecki suggest that "GATT-type emergency protection is difficult to rationalize in the services context because in many cases it will require taking action against foreign firms that have established a commercial presence."[11] They question why a government would take such an action that would harm the investment environment and would negatively affect the national employees of the targeted foreign-owned firms. But government actions belie that logic, at least regarding energy. UK regulators showed little concern for the effect on foreign firms when they placed a moratorium on the construction of new gas-fired power plants in 1997.[12] The economic effects associated with a crisis or other disruption in energy markets can be large and damaging.

A precedent for safeguard measures already exists for energy in the context of EU energy integration. For example, article 24 of the EU gas directive states that member-states may "temporarily take the necessary safeguard measures" during a sudden crisis in the energy market, when the system integrity is threatened.[13] Because those measures came into effect after the UK moratorium on gas-fired power plants, whether that action could have been justified as a safeguard measure is unclear. The inclusion of the safeguard instruments in the directive, however, is indicative of the special economic and security sensitivities associated with energy.

Incorporating a safety valve within the WTO agreement on energy services may be useful in securing deeper commitments from countries. Previous service negotiations demonstrate that countries sometimes make binding commitments below their existing levels of liberalization.[14] Governments face considerable uncertainty regarding liberalization. They can never fully predict demand for more protection at home. Governments respond to the uncertainty by seeking flexibility.[15] Though problematic from the perspective of liberalization, such behavior can be perfectly rational. Governments compensate for the lack of flexibility by undercommitting.

Given those considerations, governments may find the inclusion of emergency safeguard instruments in the negotiations on energy services worthwhile. That inclusion, however, should be temporary and applied only to the extent necessary to prevent or remedy serious injury, as the safeguards are in the context of goods. Provided they are framed to minimize abuse, emergency safeguards may elicit more meaningful commitments from governments than can be otherwise expected.

Government Procurement

Government procurement, a final area that trade negotiators must consider, can be a catalyst in stimulating competition. By opening up procurement to all competitors, policies on government procurement can help establish new standards for entry and more competitive market conditions. More often, however, government policies restrict competition to preferred—generally domestic—suppliers. The policies may be legislated, as are U.S. buy-America policies, which discriminate against foreign suppliers. Or policies—often those dealing with procurement of oil field services—may be determined on an administrative or political basis with little transparency or "challenge procedures" that give firms an opportunity to object before the final procurement decision is made.

The significance of discriminatory procurement practices depends on several factors. One is the size of government procurement relative to the market. In some cases governments may play a minor role in service markets but a major role in others, particularly if a government body holds a monopoly position as many power and gas utilities do. Another factor is how tradable a product is. Government procurement policies are unlikely to have a major effect on the price of products traded in world markets, such as oil. Many services and locally produced goods, such as electricity, however, are less likely to be traded. A third factor relates to the contestableness of the market. Discriminatory practices are less likely to affect markets primarily supplied by private firms backed by strong pro-competition policies.

The GATS does not now cover services supplied in the "exercise of governmental authority." The lack of coverage, however, has not convinced everyone of a necessity for stand-alone disciplines on government procurement. Simon Evenett and Bernard Hoekman argue that there is no compelling reason to treat the procurement of goods and services differently and that countries should instead focus on developing broad horizontal disciplines.[16] They suggest that expanding commitments regarding market access and national treatment under the GATS will lessen any need for multilateral rules on procurement. Their suggested focus is removing barriers to market access, such as the right of establishment, and cross-cutting transparency disciplines. But governments and other stakeholders do need to determine whether government procurement of energy is sufficiently great and market distorting to warrant the development of specific rules for energy services.

7

Conclusion

The GATS provides a framework for countries to expand their commitments significantly to market access and national treatment in the area of energy services. It also provides a way for countries to make additional commitments in important areas such as third-party access to essential facilities, competition, and independent regulation. Achieving those goals requires clarifying the existing classification system and making it more relevant to the way in which energy is produced and delivered. It will also require that countries not only make binding commitments but follow through by putting in place the regulatory structures and procedures necessary to meet their obligations and specific commitments according to the negotiated deadlines.

Securing those outcomes will be challenging, given the economic importance, complexity, and political sensitivity of the energy sector. But reasons for optimism exist. The era of vertically integrated monopolies with clearly defined service territories and locked-in customer bases is giving way to more flexible market arrangements. The competitive transformation of the energy industry holds the potential to meet the demand for energy more efficiently and cost-effectively. It also holds the potential for greater innovation in the way energy services are bundled and delivered, which can yield cleaner, more reliable, and more reasonably priced energy. Increasingly, developed and developing countries are recognizing that the benefits of market allocation depend on establishing fair and effective administrative rules and regulations not only in the domestic context but also for international trade. That recognition is an important factor behind the recent interest in bringing GATS disciplines to bear more effectively in the energy sector.

The decision reached at the ministerial conference in Doha, Qatar, in November 2001 to launch a new round of global trade talks was a significant development for negotiations on energy services. Discussions on energy services had been taken up as part of the unfinished business of the Uruguay Round, but progress was slow. Folding energy services into the broader Doha agenda has created new momentum. A broader array of issues sits on the table for linkage and exchange between countries. A specific timetable must be met. WTO members were to submit requests for market access by June 30, 2002, followed by initial offers of market access by March 2003. Final commitments are expected no later than January 1, 2005. Taking advantage of that opportunity to reach a global trade agreement on energy services will help to ensure that developed and developing countries reap the full benefits of more open and competitive energy markets.

Notes

Chapter 1: Introduction

1. Calculated from data reported in British Petroleum, *BP Statistical Review of World Energy,* June 2001, at http://www.bp.com/centres/energy/index.asp, and U.S. Department of Energy, Energy Information Administration, *International Energy Outlook 2001* (Washington, D.C.: Government Printing Office, 2001).
2. Organization for Economic Cooperation and Development, International Energy Agency, *World Energy Outlook 2000* (Paris: IEA/OECD, 2000), pp. 47–54. Developing regions include China, South Asia, East Asia, Latin America, Africa, and the Middle East.

Chapter 2: Importance of Energy Services

1. See Pier Angelo Toninelli, ed., *The Rise and Fall of State-Owned Enterprise in the Western World* (New York: Cambridge University Press, 2000), and IEA, *The Role of IEA Governments in Energy* (Paris: IEA/OECD, 1996).
2. Andersen, *Global E&P Trends 2001* (Houston: Andersen, 2002).
3. The International Energy Agency projects that during the 1997–2020 period total oil supply will decline by an average annual rate of 1.4 percent among OECD countries but will increase by 1.1 percent in Russia, 3.9 percent in the OPEC Middle East, 5.3 percent in the Caspian, and 5.7 percent in Latin America. See OECD, *World Energy Outlook 2000,* p. 77.
4. Cambridge Energy Research Associates, *Global Power Horizons: Strategic and Regional Outlooks* (Cambridge: 2001).
5. Cambridge Energy Research Associates, *Electric Power Trends 2001* (Cambridge, Mass.: 2000), p. 73.
6. *OECD, World Energy Outlook 2000,* p. 94.

7. "Kenya and Trade in Energy Services in East Africa," paper presented at "Experts Meeting on Energy Services in International Trade: Development Implications," Geneva, UN Conference on Trade and Development, 2001.

8. Patricia L. Williams, "Bill Outsourcing: What's the Advantage?" *Public Utilities Fortnightly,* IT suppl., July 2000, pp. 55–66.

9. Melanie Mauldin, "Information: The Key to Unlocking Value from Customer Choice," in *Customer Choice: Finding Value in Retail Electricity Markets,* edited by Ahmad Faruqui and J. Robert Malko (Vienna, Va.: Public Utilities Reports, 1999), pp. 263–86.

10. IEA, *Energy Policies of IEA Countries: 2000 Review* (Paris: IEA/OECD, 2001), p. 47.

11. Ibid.

12. Companies providing integrated energy services include, for utilities, AEP Energy Services, CMS Energy Corp., Duke Energy Corp., Electricité de France, Eon, and Sempra Energy; for oil, Royal Dutch/Shell, BP, and Norsk Hydro; and for manufacturing, General Electric, ABB, Honeywell, Johnson Controls, and Siemens.

Chapter 3: Liberalizing Energy Markets

1. Joseph P. Kalt, *The Economics and Politics of Oil Price Regulation: Federal Policy in the Post-Embargo Era* (Cambridge: MIT Press, 1981), pp. 233–34.

2. A. F. Alhajji and David Huettner, "OPEC and Other Commodity Cartels: A Comparison," *Energy Policy* 28 (2000): 1161.

3. Kosuke Oyama, "The Policy Making Process behind Petroleum Industry Regulatory Reform," in *Is Japan Really Changing Its Ways? Regulatory Reform and the Japanese Economy,* edited by Lonny E. Carlile and Mark C. Tilton (Washington, D.C.: Brookings Institution, 1998), p. 147.

4. Bruce Henning, Lee Tucker, and Cindy Liu, "Productivity Improvements in the Natural Gas Distribution and Transmission Industry," *Gas Energy Review* 23 (February 1995): 17–20.

5. Paul W. MacAvoy, *The Natural Gas Market: Sixty Years of Regulation and Deregulation* (New Haven: Yale University Press, 2000), p. 81.

6. Commission of the European Communities, "Completing the Internal Energy Market," Commission staff working paper, Brussels, March 12, 2001, pp. 24–25.

7. Ibid., pp. 18–24.

8. Faye Steiner, "Regulation, Industry Structure and Performance in the Electricity Supply Industry," *OECD Economic Studies* 32, January 2001.

9. Samantha Doove, Own Gabbitas, Du Nguyen-Hong, and Joe Owen, *Price Effects of Regulation: International Air Passenger Transport, Telecommunications and Electricity Supply,* Productivity Commission staff research paper, AusInfo, Canberra, October 2001.

10. Paul L. Joskow, "Deregulation and Regulatory Reform in the U.S. Electric Power Sector," in *Deregulation of Network Industries: What's Next?* edited by Sam Peltzman and Clifford Winston (Washington, D.C.: Brookings Institution, 2000), p. 121.

11. See, for example, Severin Borenstein, "The Trouble with Electricity Markets: Understanding California's Restructuring Disaster," *Journal of Economic Perspectives* 16 (1) (winter 2002): 191–211.

12. See Pat Wood, chairman, Federal Energy Regulatory Commission, testimony, U.S. Senate Committee on Energy and Natural Resources, January 29, 2002, at http://www.ferc.gov/news/congressional testimony/testimony.htm#enron.

13. For detailed explanations of the barriers in selected markets, see CEC, *European Overview,* vol. 1 of *Report to the European Commission Directorate General for Transport and Energy to Determine Changes after Opening of the Gas Market in August 2000,* DRI-WEFA (Brussels: CEC, 2001), and U.S. International Trade Commission, *Electric Power Services: Recent Reforms in Selected Foreign Markets,* Investigation 332–441, USITC publication 3370 (Washington, D.C.: Government Printing Office, 2000).

14. Jonathan P. Stern, *Competition and Liberalization in European Gas Markets* (Washington, D.C.: Brookings Institution, 1998), and Jacques Percebois, "The Gas Deregulation Process in Europe: Economic and Political Approach," *Energy Policy* 27 (1999): 9–15.

15. As part of its annual trade review with Japan, the U.S. government has raised the issue of establishing access for all market participants but has stopped short of demanding mandatory access with regulated tariffs for competing suppliers. See U.S. Trade Representative, "Annual Reform Recommendations from the Government of the United States to the Government of Japan under the U.S.-Japan Regulatory Reform and Competition Policy Initiative," October 14, 2001, annex 11–15.

Chapter 4: Energy Services and the GATS

1. In 1994 more than sixty WTO member countries submitted MFN exemptions, concentrated in three sectors: audiovisual, financial services, and transportation (road, air, and maritime).

2. Rachel Thompson and Keiya Iida, "Strengthening Regulatory Transparency: Insights for the GATS from the Regulatory Reform Country Reviews," in *Trade in Services: Negotiating Issues and Approaches* (Paris: Organization for Economic Cooperation and Development, 2001), p. 112.

3. Keiya Iida and Julia Nielson, "Transparency in Domestic Regulation: Prior Consultation," in *Trade in Services: Negotiating Issues and Approaches, Industry, Services and Trade* (Paris: OECD, 2001), pp. 115–35.

4. Christopher Melly, "Power Market Reform and International Trade in Services," paper presented at "Experts Meeting on Energy Services in International Trade: Development Implications," Geneva, UNCTAD, 2001.

5. For an extended discussion of these commitments, see "Energy Services: Background Note by the Secretariat," Council for Trade in Services, World Trade Organization, S/C/W/52, September 9, 1998, pp. 19–35.

6. "Guidelines for the Scheduling of Specific Commitments under the General Agreement on Trade in Services (GATS)," World Trade Organization, S/L/92, March 28, 2001, p. 8.

7. For a more complete description of the approach, see Rachel Thompson, "Integrating Energy Services into the World Trading System," Energy Services Coalition, Washington, D.C., April 10, 2000.

8. Personal communication, U.S. Trade Representative staff, April 11, 2002.

9. "Services Sectoral Classification List: Note by the Secretariat," World Trade Organization, MTN.GNS/W/120, July 10, 1991, p. 3.

Chapter 5: Necessary Additional Commitments

1. Aaditya Mattoo, "Shaping Future GATS Rules for Trade in Services," Development Research Group, World Bank, Washington, D.C., June 2000, p. 12.

2. See Daniel Roseman, "Domestic Regulation and Trade in Telecommunications Services: Experience and Prospects under the GATS," paper delivered at OECD–World Bank Services Experts Meeting, OECD, Paris, March 4–5, 2002.

3. Various U.S. court cases spell out useful criteria for determining the reasonableness of a refusal to deal. See in particular MCI Communications Corp. v. American Tel. & Tel., 708 F.2d 1081 (7th cir., 1983).

4. For more on how prior consultation might work, see Iida and Nielson, "Transparency in Domestic Regulation," pp. 115–35.

5. Steven R. Salbu, "Battling Global Corruption in the New Millennium," *Law and Policy in International Business* 31 (1) (1999): 47–78.

6. Catherine D. Wolfram, "Measuring Duopoly in the British Electricity Spot Market," *American Economic Review* 89 (4) (September 1999): 821.

7. Pablo T. Spiller, "Regulatory Commitment and Utilities' Privatization: Implications for Future Comparative Research," in *Modern Political Economy: Old Topics, New Directions,* edited by Jeffrey S. Banks and Eric A. Hanushek (New York: Cambridge University Press, 1997), p. 64.

8. See IEA, *Regulatory Institutions in Liberalized Electricity Markets* (Paris: OECD, 2001), and "Power and Gas Regulation—Issues and International Experience," draft working paper, World Bank, Washington, D.C., April 2001.

Chapter 6: Additional Issues

1. Some argue that too much flexibility in the GATS framework makes it a relatively weak instrument in advancing liberalization. See Patrick Low and Aaditya Mattoo, "Is There a Better Way? Alternative Approaches to Liberalization under the GATS," in *GATS 2000: New Directions in Services Trade Liberalization,* edited by Pierre Sauvé and Robert Stern (Washington, D.C.: Brookings Institution, 2000), pp. 449–72.

2. United Nations, *World Investment Report 2001: Promoting Linkages,* United Nations Conference on Trade and Investment, United Nations (New York: UN, 2001) (internet edition), pp. 16–17, at http://www.unctad.org/en/pub/pubframe.htm.

3. Communication from Venezuela, "Negotiating Proposal on Energy Services," Council for Trade in Services, Special Session, World Trade Organization, S/CSS/W/69, March 29, 2001, sect. 3, 15.

4. Bijit Bora, Peter J. Lloyd, and Mari Pangestu, "Industrial Policy and the WTO," *World Economy* 23 (4) (April 2000): 554–55.

5. J. Gregory Sidak and Daniel F. Spulber, "Deregulation and Managed Competition in Network Industries," *Yale Journal on Regulation* 15 (1) (winter 1998): 146.

6. See, for example, Tomoyuki Takao, "TEPCO's Position on WTO Energy Services Trade Negotiations," Tokyo Electric Power Company, Corporate Planning Department, September 2001.

7. Simone Espey, "Renewables Portfolio Standards: A Means for Trade with Electricity from Renewable Energy Sources?" *Energy Policy* 29 (2001): 557–66.

8. Aaditya Mattoo, "Shaping Future GATS Rules for Trade in Services," p. 9.

9. Communication from the United States, "Framework for Negotiation," Council for Trade in Services, special session, World Trade Organization, S/CSS/W/4, July 13, 2000, sect. 2, and communication from Venezuela, "Negotiating Proposal on Energy Services," Council for Trade in Services, special session, World Trade Organization, S/CSS/W/69, March 29, 2001, sect. 3, 16.

10. Communication from the EC and member-states, "GATS 2000: Energy Services," Council for Trade in Services, special session, World Trade Organization, S/CSS/W/60, March 22, 2001, sect. 1, 5, and communication from Japan, "Negotiation Proposal on Energy Services," Council for Trade in Services, special session, World Trade Organization, S/CSS/W/42/suppl. 3, October 4, 2001, sect. 1, 5.

11. Bernard M. Hoekman and Michel M. Kostecki, *The Political Economy of the World Trading System: The WTO and Beyond,* 2nd ed. (New York: Oxford University Press), p. 270.

12. Peter C. Evans, "Energy Services, Domestic Regulation and the WTO," OECD/World Bank Conference on Regulatory Reform and Trade Liberalization in Services, Paris, March 4–5, 2002.

13. Directive 98/30/EC of the European Parliament and of the Council, June 22, 1998, "Concerning Common Rules for the Internal Market in Natural Gas," *Official Journal* L204 (July 21, 1998): 1–12.

14. One example is the level of foreign equity participation permitted in commercial banks. The Philippine government made a binding commitment of only 51 percent even though domestic law allows 60 percent. See Aaditya Mattoo, "Financial Services and the WTO: Liberalization Commitments of the Developing and Transition Economies," *World Economy* 23 (3) (March 2000): 371.

15. B. Peter Rosendorff and Helen V. Milner, "The Optimal Design of International Trade Institutions: Uncertainty and Escape," *International Organization* 55 (4) (autumn 2001): 831.

16. Simon J. Evenett and Bernard M. Hoekman, "Government Procurement of Services and Multilateral Disciplines," in *GATS 2000: New Directions in Services Trade Liberalization,* edited by Pierre Sauvé and Robert Stern (Washington, D.C.: Brookings Institution, 2000), pp. 143–63.

About the Author

PETER C. EVANS specializes in international political economy, energy market liberalization, and international trade. He was a senior associate at Cambridge Energy Research Associates and has been a consultant to the World Bank, U.S. Department of Energy, and private energy and technology companies in Australia, Japan, and the United States. He was a visiting scholar at the Central Research Institute for Electric Power Industry, Tokyo, Japan, from 1991 to 1993.

Recent publications include the chapter "Meeting the International Competition: Conflict and Cooperation in Export Financing," with Kenneth Oye, in *U.S. Ex-Im Bank in the Twenty-first Century* (Institute for International Economics, January 2001).

Mr. Evans holds a BA from Hampshire College and an MS from the Massachusetts Institute of Technology, where he is completing his PhD in the Department of Political Science.

Daniel Patrick Moynihan
University Professor
Maxwell School
Syracuse University

Sam Peltzman
Ralph and Dorothy Keller
Distinguished Service Professor
of Economics
University of Chicago
Graduate School of Business

Nelson W. Polsby
Heller Professor of Political Science
Institute of Government Studies
University of California, Berkeley

George L. Priest
John M. Olin Professor of Law and
Economics
Yale Law School

Jeremy Rabkin
Professor of Government
Cornell University

Murray L. Weidenbaum
Mallinckrodt Distinguished
University Professor
Washington University

Richard J. Zeckhauser
Frank Plumpton Ramsey Professor
of Political Economy
Kennedy School of Government
Harvard University

Research Staff

Joseph Antos
Resident Scholar

Leon Aron
Resident Scholar

Claude E. Barfield
Resident Scholar; Director, Science
and Technology Policy Studies

Walter Berns
Resident Scholar

Douglas J. Besharov
Joseph J. and Violet Jacobs
Scholar in Social Welfare Studies

Robert H. Bork
Senior Fellow

Karlyn H. Bowman
Resident Fellow

John E. Calfee
Resident Scholar

Charles W. Calomiris
Arthur F. Burns Scholar in
Economics

Lynne V. Cheney
Senior Fellow

Nicholas Eberstadt
Henry Wendt Scholar in Political
Economy

Eric M. Engen
Resident Scholar

Mark Falcoff
Resident Scholar

J. Michael Finger
Resident Fellow

Gerald R. Ford
Distinguished Fellow

Murray F. Foss
Visiting Scholar

David Frum
Visiting Fellow

Harold Furchtgott-Roth
Visiting Fellow

Reuel Marc Gerecht
Resident Fellow

Newt Gingrich
Senior Fellow

James K. Glassman
Resident Fellow

Robert A. Goldwin
Resident Scholar

Scott Gottlieb
Resident Fellow

Michael S. Greve
John G. Searle Scholar

Robert W. Hahn
Resident Scholar; Director,
AEI-Brookings Joint Center
for Regulatory Studies

Kevin A. Hassett
Resident Scholar

Steven F. Hayward
F. K. Weyerhaeuser Fellow

Robert B. Helms
Resident Scholar; Director, Health
Policy Studies

Leon R. Kass
Hertog Fellow

Jeane J. Kirkpatrick
Senior Fellow; Director, Foreign and
Defense Policy Studies

Marvin H. Kosters
Resident Scholar; Director,
Economic Policy Studies

Irving Kristol
Senior Fellow

Michael A. Ledeen
Freedom Scholar

James R. Lilley
Resident Fellow

John R. Lott, Jr.
Resident Scholar

Randall Lutter
Resident Scholar

John H. Makin
Resident Scholar; Director,
Fiscal Policy Studies

Allan H. Meltzer
Visiting Scholar

Joshua Muravchik
Resident Scholar

Charles Murray
Bradley Fellow

Michael Novak
George Frederick Jewett Scholar
in Religion, Philosophy, and Public
Policy; Director, Social and Political
Studies

Norman J. Ornstein
Resident Scholar

Richard Perle
Resident Fellow

Sarath Rajapatirana
Visiting Scholar

Sally Satel
W. H. Brady, Jr., Fellow

William Schneider
Resident Fellow

J. Gregory Sidak
Resident Scholar

Radek Sikorski
Resident Fellow; Executive
Director, New Atlantic Initiative

Christina Hoff Sommers
Resident Scholar

Arthur Waldron
Visiting Scholar; Director, Asian
Studies

Graham Walker
Visiting Scholar

Peter J. Wallison
Resident Fellow

Ben J. Wattenberg
Senior Fellow

Karl Zinsmeister
J. B. Fuqua Fellow; Editor,
The American Enterprise